# VALVERDE'S GOLD

The Royal Geographical Society
Llanganati Par

Steven J. Charbonneau

VALVERDE'S GOLD: THE ROYAL GEORAPHICAL SOCIETY LLANGANATI PAPERS. Copyright © 2012 by Steven J. Charbonneau. All rights reserved. No part of this book may be used or reproduced in any manner whatsoever without written permission except in the case of brief quotations embodied in critical articles or reviews. Portions originally published by The Royal Geographical Society of London are currently within the public domain. Compiled and written by Steven J. Charbonneau. Printed in The United States of America.

Please direct corrections, comments, questions or requests to:
lustforincagold @ yahoo.com

Second Paperback Edition
ISBN - 13: 978-1479240555

# TABLE OF CONTENTS

### INTRODUCTION

Page 4

### ON THE MOUNTAINS OF LLANGANATI IN THE EASTERN CORDILLERA OF THE QUITONIAN ANDES

Page 23

### TRAVELS IN ECUADOR

Page 71

### THE INCA TREASURE OF LLANGANATI

Page 98

### EPILOGUE

Page 105

# INTRODUCTION

Although Christopher Columbus was not the first European explorer to have ever reached the Americas, his four round trip voyages between 1492 and 1503 led to the first lasting European contact with the Americas. Columbus's discoveries inaugurated a period of Spanish exploration and conquest of this "New World" providing great benefit to the Kingdom of Spain, and even greater detriment to the indigenous peoples and their cultures!

In 1513 Vasco Núñez de Balboa was the first to cross the Isthmus of Panama in search of gold, but instead discovered the Mar del Sur, Pacific Ocean or South Sea, and in turn the west coast of this "New World". Balboa claimed the Pacific Ocean and all the lands adjoining it for the Spanish Crown.

The age of discovery and conquest of the Americas commenced in earnest with the destruction and conquest of the Aztec Empire of Mexico by Hernán Cortés between 1519 and 1521. There was a difference though between the Spanish conquest of Mexico (the Yucatán) and their conquest of the Aztec Empire. The Aztec Empire was a conquest by campaign, while the conquest of the Yucatán was a much longer campaign, lasting from 1551 to 1697.

The Maya peoples of the Mayan civilization in the Yucatán Peninsula were indigenous to present day Mexico and northern Central America. Unlike the Aztec and Inca Empires, there was no single Maya political center that once overthrown would precipitate the end of resistance from the indigenous peoples. Instead, the conquistadors were required to subdue the numerous independent Maya states and rulers one by one, many of which kept up a fierce resistance until the bitter end.

The Spanish soon discovered that the Mayan lands of the Yucatán Peninsula did not possess the great wealth to be had from the seizure of precious metal resources such as gold or silver. Considering that most of the conquistadors had been motivated by the prospects of this wealth, expeditions set out in search of "El Dorado" a mysterious land called "Peru," rumored to be awash in gold and silver!

Spanish conquistador Francisco Pizarro commenced his first voyage of three, south along the Pacific Coast of the "New World" in 1524. Pizarro's quest would culminate at the end of 1532 with the capture of the Inca Atahualpa, whose

Realm encompassed some 300,000 square miles, the majority of modern day Columbia, Ecuador, Peru, Bolivia, Argentina and Chile!

Atahualpa's capture led to an unfathomable ransom of over twenty tons of gold and silver being paid for his release! In place of being released Atahualpa was accused of a crime, found guilty and murdered. The consequences of this action were that untold quantities of gold and silver destined for the ransom of Atahualpa were secreted throughout the Realm! Even so, the conquistadores then advanced on the Incan southern capital of Cuzco, where they looted yet another twenty tons of Incan gold and silver!

The transport of treasure and perhaps Atahualpa's mummy into the Llanganatis would have been lost to memory, myth and legend, if not for the involvement of a young Spaniard named Valverde, his Indian wife, the King of Spain and Richard Spruce!

[]

The Spanish chroniclers and countless historians have left a wealth of information and knowledge about this great empire, its peoples and their conquest; *The Incas of Pedro de Cieza de León, Reports on the Discovery of Peru, History of the Conquest of Peru* (1847), and *The Conquest of the Incas* (1970) to name but a few. My brief work entitled *Inca Gold: History, Conquest & Legend* (2012) also provides pertinent historical background concerning the Royal Geographical Society

Llanganati papers published herein, written by impeccable sources . . . men of great character and fortitude.

[]

The Royal Geographical Society of England is the learned society and professional body for geography and geographers. The society was founded in 1830 under the name Geographical Society of London, as an institution to promote the "advancement of geographical science." Like many learned societies, it had started as a dining club in London, where select members held informal dinner debates on current scientific issues and ideas. Under the patronage of King William IV the society later became known as The Royal Geographical Society and was granted its Royal Charter under Queen Victoria in 1859.

The early history of the Society was interlinked with the history of British Geography, exploration and discovery. Information, maps, charts and knowledge gathered on expeditions were either deposited with, or presented to, the Royal Geographical Society through lectures given at the societies meetings, making up its now unique geographical collections.

[]

Richard Spruce Esq. [1817-1893], named after his father and grandfather, was born in September 1817 at Ganthorpe near Terrington, in the West Midland region of England, the only child of Richard and Ann. Richard's mother having died, his father was remarried to Mary Priest shortly thereafter.

Spruce took lessons in Latin and Greek from a retired schoolmaster, taught himself to read and write French, and in his latter years acquired Portuguese, Spanish and three Indian languages including Quichua. It appears however that Richard was mainly educated by his father who was a respected schoolmaster.

Spruce had developed a great love of nature at an early age and had taken up botany by the 1830's and by 1834, when merely sixteen years old, had produced a list of 403 species that he had found around Ganthorpe. In 1837 Spruce had drawn up yet another *List of the Flora of the Malton District* containing 485 species of flowering plants.

Following his father's footsteps into teaching, Spruce obtained the position of mathematical master at the Collegiate School of York at the end of 1839, which he retained until the school was closed in mid 1844. Spruce's teaching position had provided him with the benefits of a regular salary and spare time, which allowed him to continue and explore many parts of York where he made numerous botanical discoveries.

Once Spruce found himself without a job, he became determined to find employment as a botanist. In the end

Spruce decided to go to the French Pyrenees. Spruce left England in May 1845 and returned in April of the following year. Spruce had collected more than three hundred species of plants and mosses, of which seventeen were new to science and many not previously recorded from the Pyrenees.

After some consideration Spruce later became determined to conduct a botanical exploration of the Amazon. Spruce finally left England in June 1849 and arrived at Pará in the northern part of Brazil in July. Afterwards Spruce spent three months working in the forests of the area. Spruce went further afield mapping and collecting in the basin of the previously unexplored Río Trombetas. Spruce also explored and mapped the Río Negro and some of its tributaries. In 1855 Spruce explored some of the Amazon headwaters, the Huallaga, Pastaza and Bombonasa Rivers which were particularly difficult and dangerous to navigate.

In 1857 Spruce left Tarapotó, Peru, bound for Canelos, Ecuador, ultimately arriving by foot at the town of Baños in the Andes of Ecuador, where he stayed for six months before heading to Ambato in January of 1858. Spruce found Ecuador to be in a revolutionary state during this entire period which limited his travels.

In Ecuador throughout 1860, Spruce collected seeds and living plants of the Peruvian Cinchona tree for the Government of India because they were concerned about the supply of quinine, which was essential for safeguarding the health of the Indian army. Spruce collected these plants in the rain forest below the Chimborazo volcano, packed and dispatched the material, which was then established in southern India.

While at Ambato, Spruce was suffering from the effects of rheumatic fever and also suffered a stroke and found himself partly paralyzed in the neck, back and legs. For a while Spruce struggled with his collecting, but the following year he had yet another disaster, the failure of a mercantile house in Guayaquil in which Spruce had invested most of his savings, which left him almost bankrupt. After a few more years on the coast of Ecuador and in Peru, Spruce found it impossible to continue his work and returned to England in May of 1864. Spruce arrived back in England almost penniless and in very poor health due to the effects of his stroke and an intestinal disease.

Spruce however had survived all kinds of dangers, illnesses and deprivations, working for long periods alone except for his Indian assistants. His nearly fifteen-year expedition resulted in Spruce collecting more than seven thousand species of plants and fungi. Spruce's work added enormously to the scientific knowledge of Amazonian botany.

The high quality of the specimens Spruce sent to England and the scrupulous care with which they were collected, labeled and annotated was clearly evident throughout his work. Spruce's contributions involved not only botany, he added much to the scientific description of the Andean Indians and his mapping of little known parts of Amazonia was recognized by his election as a Fellow of the Royal Geographical Society in 1866.

[]

Jordan Herbert Stabler [1885-1938], was an American Citizen, the son of Jordan and Carrie Stabler. Jordan was educated at Country School at Homewood in Maryland and graduated Johns Hopkins University in 1907 with a Bachelors Degree. Jordan became the private secretary to the Honorable Henry Lane Wilson, at the time US Minister to Belgium, where he remained until the spring of 1909. In June of 1909 Stabler was appointed Secretary of the American Legation at Quito, Ecuador, where he remained through at least June of 1911.

In March of 1911 Stabler had been appointed Second Secretary of the US Embassy at Berlin, however an emergency situation arose and he was transferred to the US Legation at

Guatemala as acting Chargé d'Affairs during the absence of the US Minister. From there Stabler was appointed as Secretary of Legation at Stockholm Sweden in 1912, served as Assistant to Chief of Division for Latin-American Affairs during 1913-1914 and in 1915 under President Woodrow Wilsons administration was appointed as Chief of Division for Latin-American Affairs.

Mr. Stabler, a Fellow of the Royal Geographical Society, apparently spent the first two years of World War I in London, retired and became affiliated with the Gulf Oil Corporation in Venezuela and Europe. Jordan died at the young age of fifty-three, in Paris.

[]

Stabler's paper drew an almost immediate response from an American . . . E. C. Brooks, apparently a friend from the period of Stabler's assignment in Quito Ecuador. The Royal Geographical Society quickly published Mr. Brooks correspondence entitled *The Inca Treasure of Llanganati* in its January 1918 *Geographical Journal.*

E. C. Brooks [1860-1922], was born Edwards Cranston Brooks on 8 Nov 1860 in Portland, Oregon, to pioneer settlers Major Quincy Adams Brooks and Elizabeth Cranston, an invalid.

Young Brooks lived a life of hardships and privations, frugal, he wasted neither words nor money. A bit of a loner with few associates, Brooks made friends with the animals on his family's farm rather than the youth of the neighborhood. Brooks obituary would later describe him "with determination and an indomitable will and perseverance," being "destitute of the feeling of fear," and of an "adventurous spirit."

As a young man of eighteen Edwards entered the State University of Oregon at Eugene, yet within two years a new opportunity opened up before him. Oregon's sole Representative in Congress announced that a competitive examination for cadetship to West Point would be held roughly three hundred miles away in Portland and on a given date. As destiny would have it, Brooks arduous solitary journey through the wilderness paid off. Out of a class of thirteen competitors for the cadetship, E. C. Brooks won the appointment from Oregon to the United States Military Academy at West Point, New York.

Young Brooks prepared for and took the examination to West Point, which he passed in June of 1882, being admitted on 1 Jul 1882 to the class of 1886. Brooks an average student, ranked fortieth in his class of seventy-seven at West Point where his most notable classmate was the elected class president, John J. Pershing, who ranked thirtieth.

Upon graduation on 1 Jul 1886, Second Lieutenant E. C. Brooks had his choice of postings, the cavalry or artillery. It has been said he chose the former arm of the service because of his great fondness for horses. Brooks was assigned to the United States 8th Cavalry and served on frontier duty in Texas, North Dakota and South Dakota, where Brooks participated in both of the battles which took place near Pine Ridge in the cavalries winter campaign against hostile Sioux Indians, and his was the cavalry battalion which brought in the body of Sitting Bull!

From there Brooks military career took him back east, across country to Delaware on college detail, having been assigned to the Delaware College in Newark as "Professor of Military Science and Tactics." Brooks was then detailed to serve at Girard College, Philadelphia, Pennsylvania as "Instructor of Cadets", during which period he was promoted to First Lieutenant of Cavalry, 6th Cavalry.

Brooks and his regiment shipped out to Cuba during the Spanish-American War with Brooks serving as Regimental Adjutant while also being engaged in the campaign against Santiago Cuba and the Battle of San Juan Hill from 1 Jul to 3 Jul 1898.

Brooks was promoted Captain and Assistant Adjutant General of the United States Volunteers on 17 Sep 1898 but was already on duty as Assistant Adjutant General, US Troops Santiago, Cuba. Brooks was also on duty as Assistant

Engineer and various other local duties at Santiago and served as aide-de-camp to Brigadier General Wood U.S. Volunteers.

Brooks was promoted and assigned as Major of the 46th US Volunteer Infantry, an assignment which he declined being also Assistant Adjutant General, Department of Santiago, Cuba, 28 Sep to 30 Dec 1899 and Auditor of the Island of Cuba from 20 Apr 1900 to 6 May 1901.

Major E. C. Brooks returned to the United States and had hardly established himself in his old quarters at Fort Myer, commanding troop, when he was ordered to the Philippines with the regiment. Brooks sailed from New York on the transport "Buford" via the Suez Canal for Manila on 21 Jan 1902. Brooks remained in the Islands "commanding troop" at Vigan, Salamogue, Camp Morrison and San Mateo until 15 Mar 1904 at which time he was en route to the United States where he would be stationed at Fort Sheridan, Illinois in "command of troop" from 15 Apr until 30 Dec 1904.

Brooks had been more than once seriously ill and confined to the hospital while in the Philippines and suffered from the trouble he contracted there, which continued to affect him more or less in the years that followed. Brooks however must have considered himself very lucky. Of his seventy-seven West Point classmates only sixty-one survived, one died in the Battle of San Juan Hill, while fifteen died in the Philippines.

Brooks, not being in very good health, officially went on leave of absence on 31 Dec 1904. Brooks furlough found him visiting his parents at Port Townsend, Washington. His resignation from the army soon followed and Major Brooks official resignation was accepted 2 May1905.

Major Brooks civilian life began with traveling from his parents in Washington State to San Francisco. On 15 May 1905, he took passage by sea, bound for Panama en route to Ecuador, "to look into a business venture he had under consideration." Brooks made a "flying trip" to London, Paris, and back in connection with this new undertaking, and he settled down in Ecuador where he engaged in various business enterprises and lived for some years. It was also during his time in Ecuador that Brooks became connected with the American Bank Note Company of New York City, where he resided upon his return from Ecuador, until his death on 14 Jan 1922.

[]

A reproduction of Don Atanasio Guzmán's map that was utilized to illustrate Spruce's original paper is presented on the following six pages. The overlapping image sequence represents portions of the map commencing at top left, middle and right, and then bottom left, middle and right.

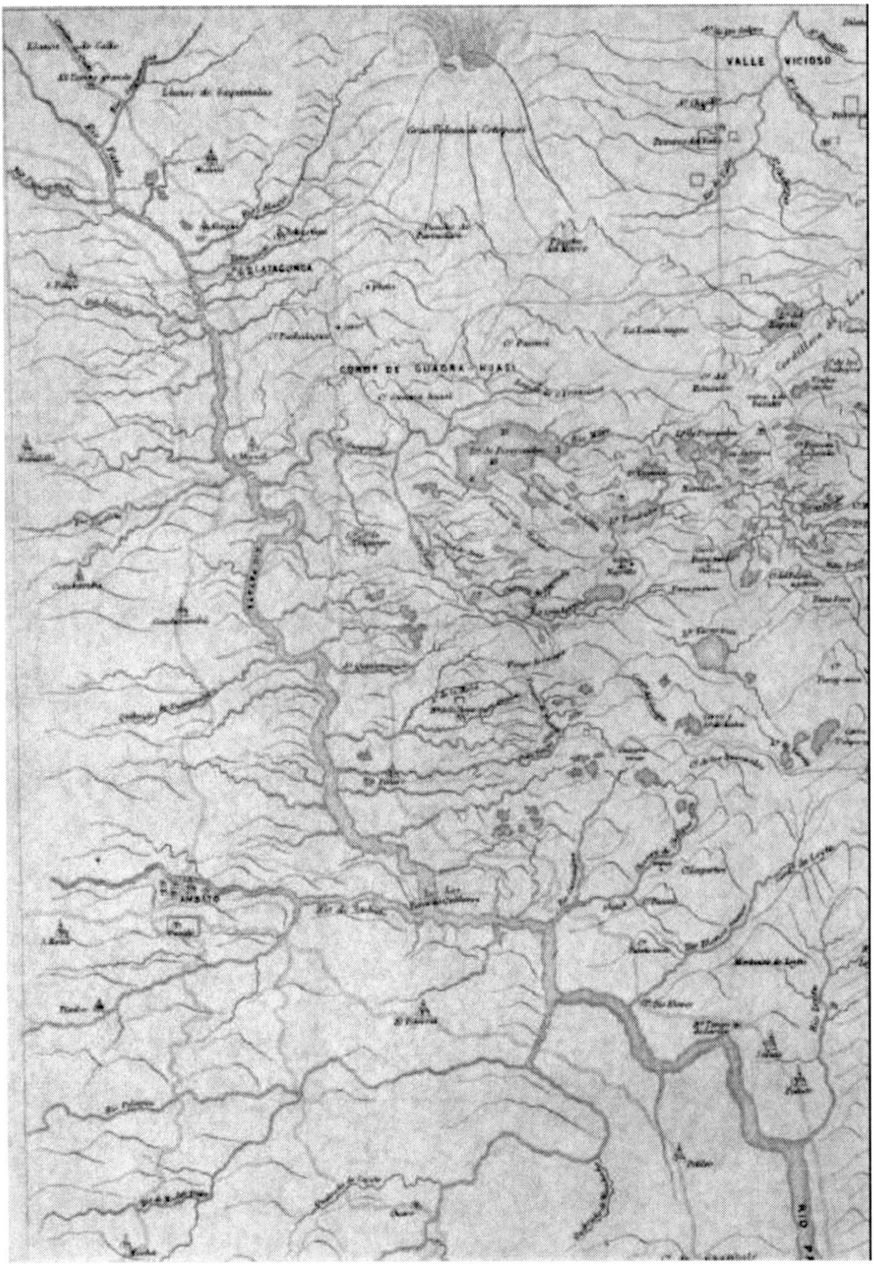
VALLE VICIOSO

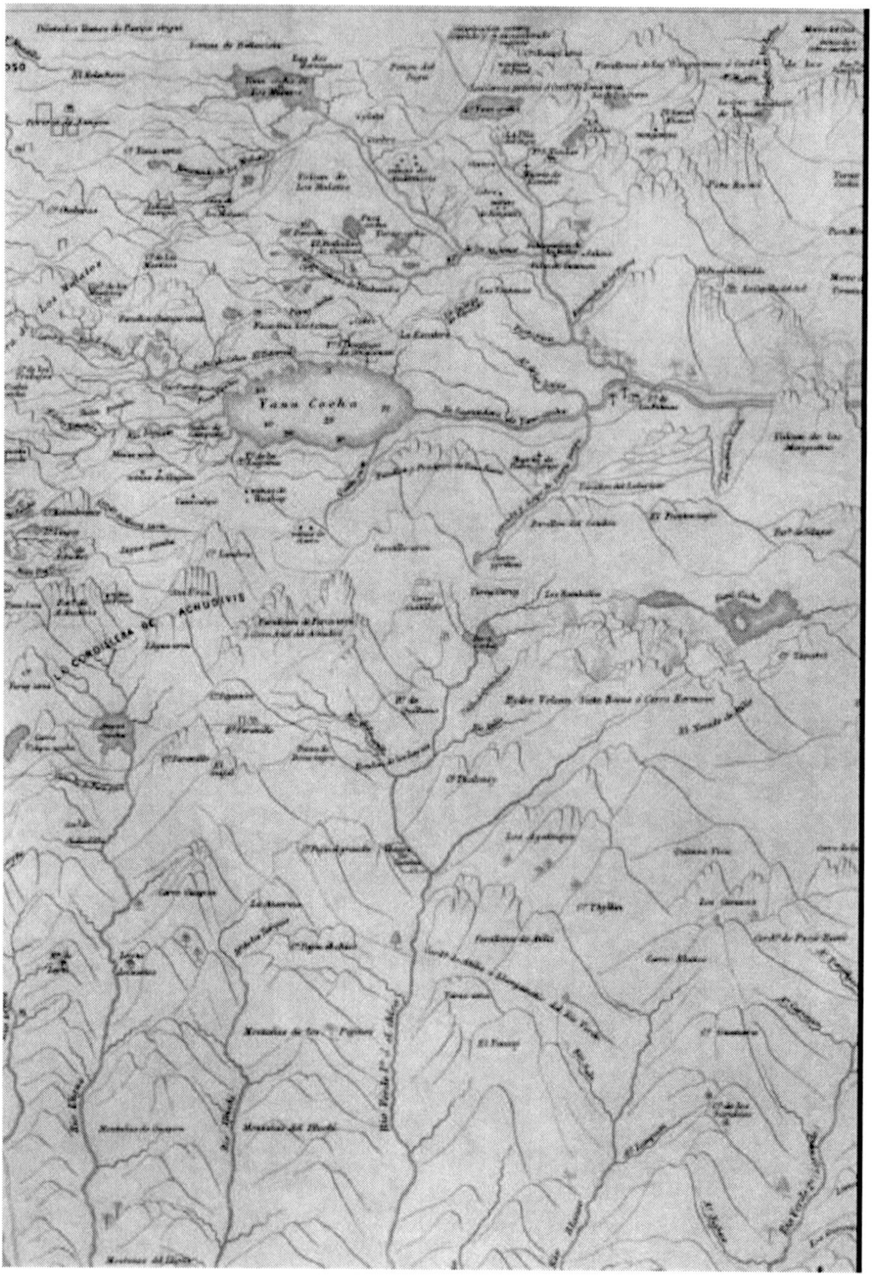
Yana Cocha

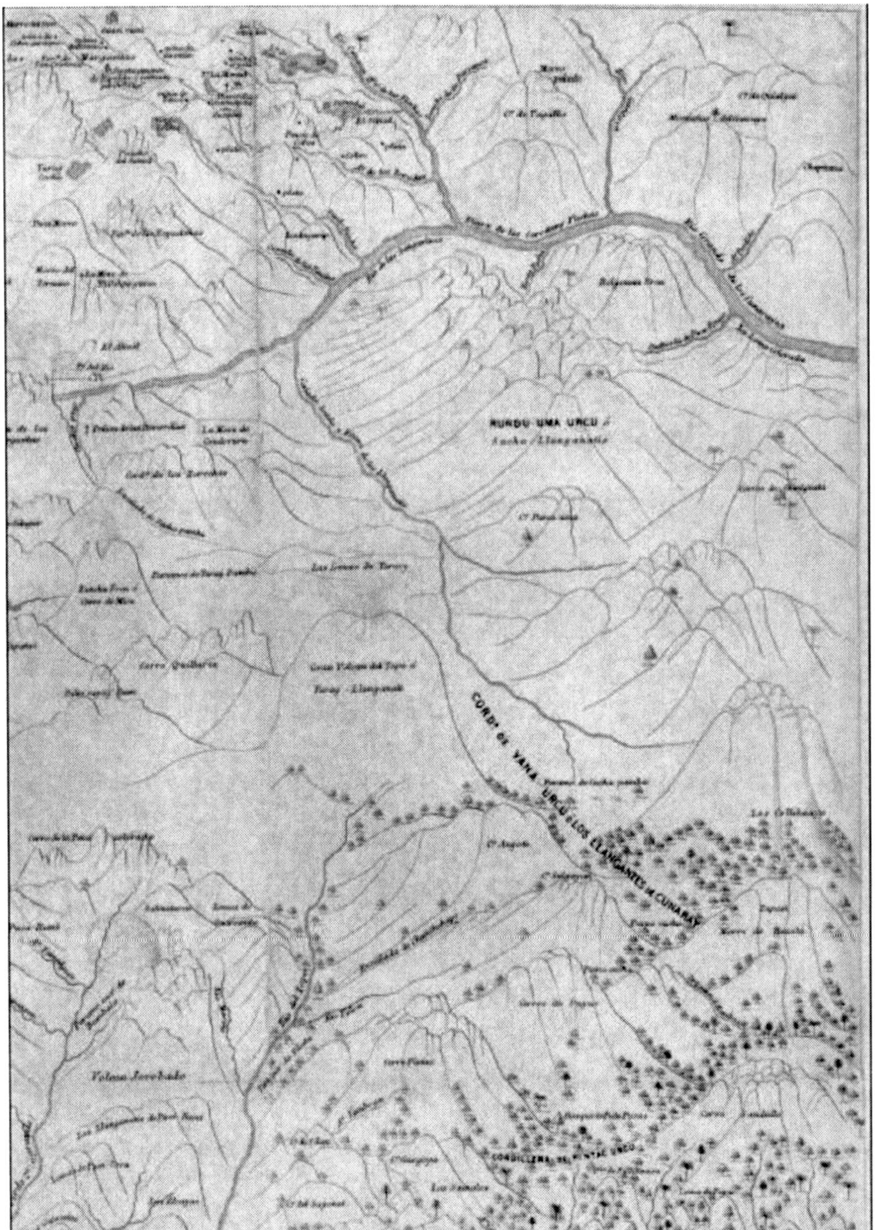
RUNDU-UMA URCU
CORDª DE YANA-URCU Ó LOS LLANGANATES Ó CUNARAY

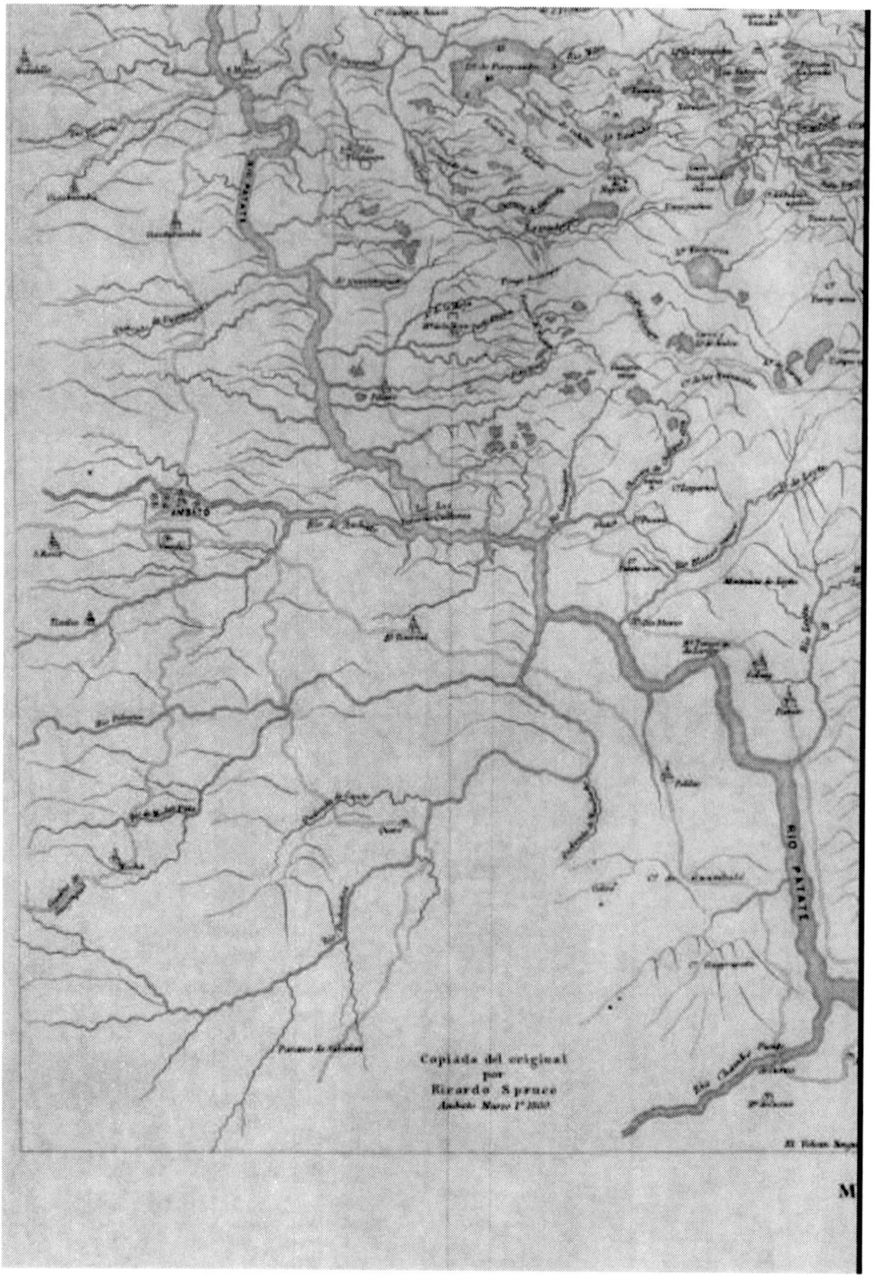


M

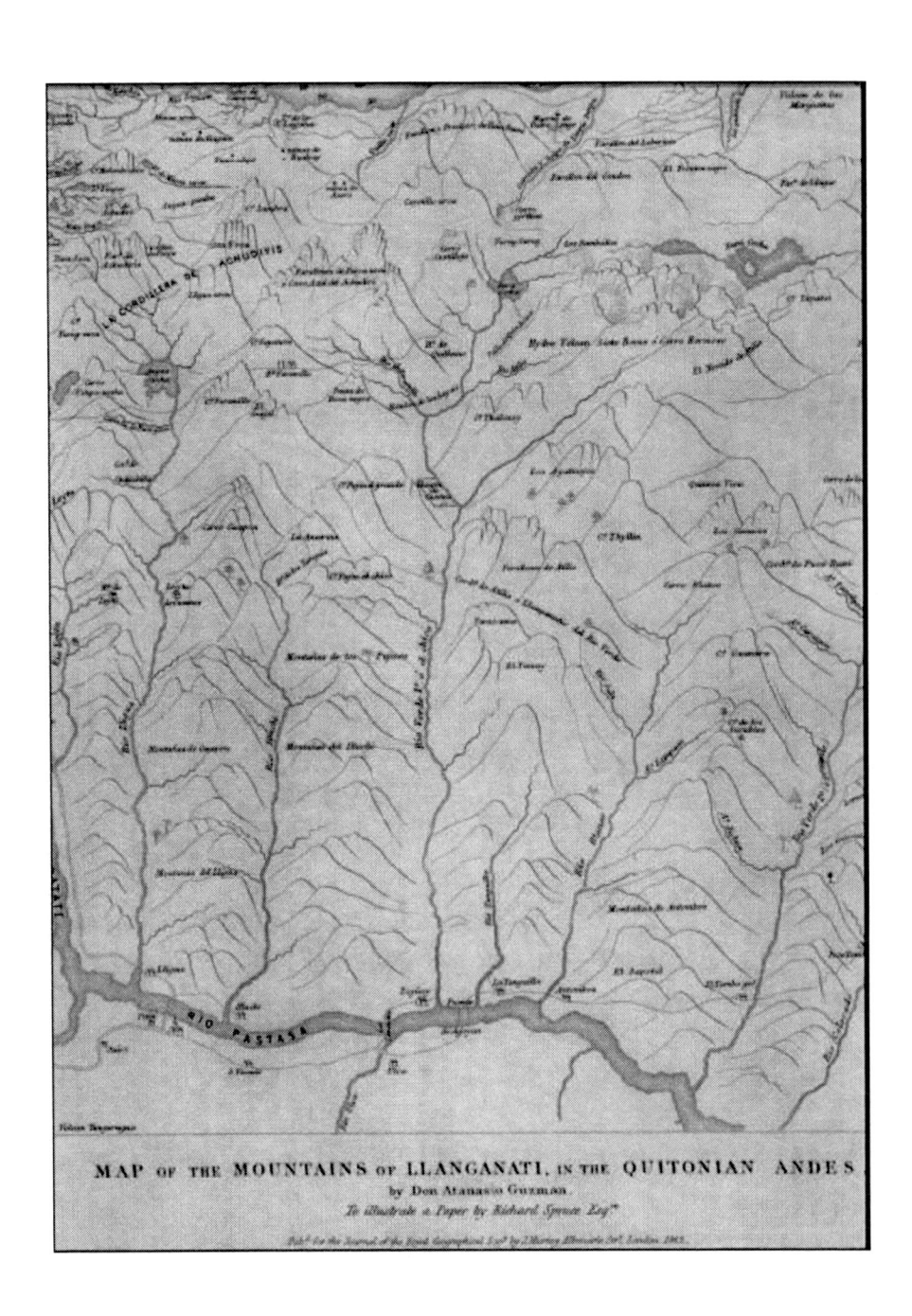

MAP OF THE MOUNTAINS OF LLANGANATI, IN THE QUITONIAN ANDES
by Don Atanasio Guzman.
To illustrate a Paper by Richard Spruce Esq.

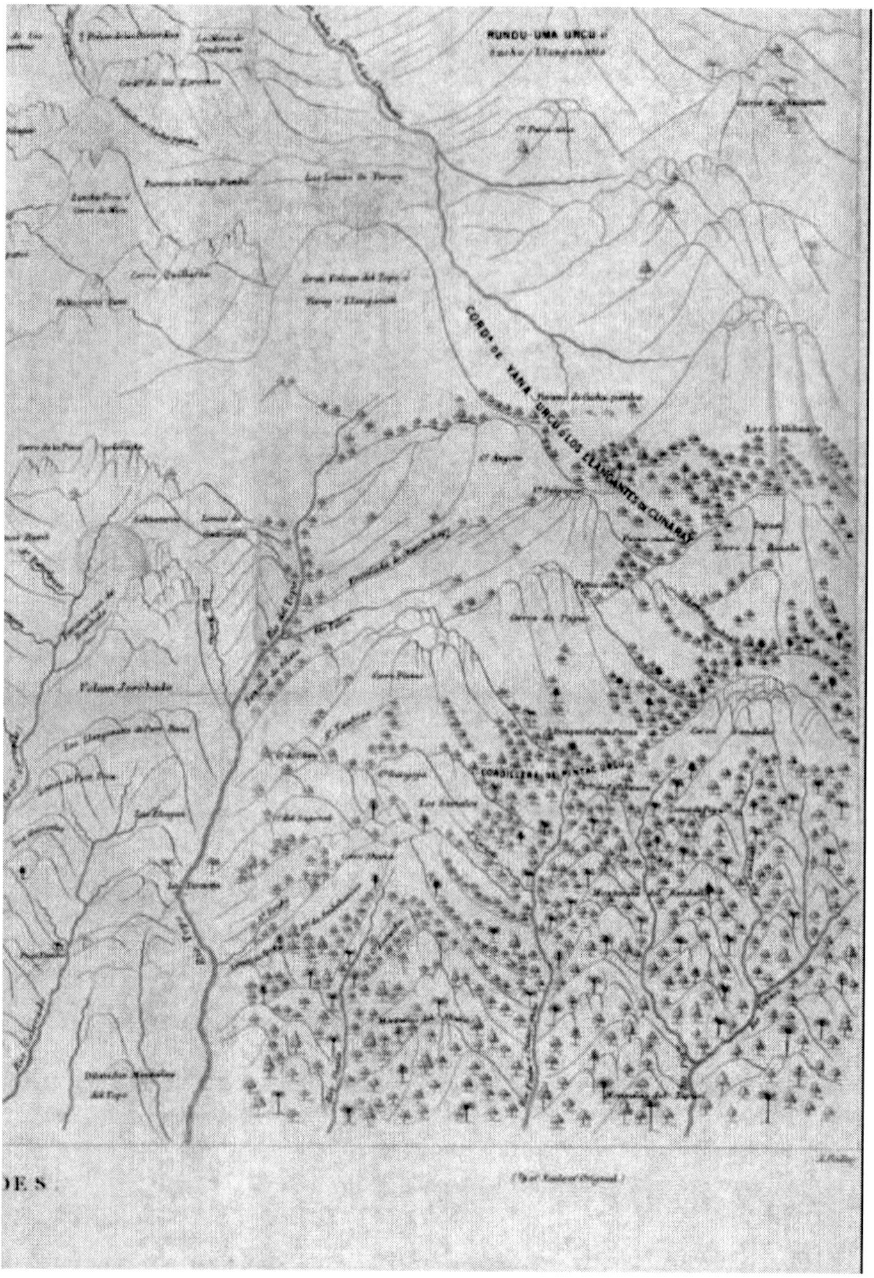

RUNDU-UMA URCU ó
CORDa DE YANA-URCU ó LOS LLANGANATES de CUNARAY
DE S.

# On the Mountains of Llanganati, in the Eastern Cordillera of the Quitonian Andes

By: RICHARD SPRUCE, Esq.

Paper originally published in:
*Proceedings of the Royal Society of London 1861*

In the year 1857 I travelled from Tarapoto, in Peru, to Baños, in Ecuador, along the rivers Huallaga, Marañon, Pastasa, and Bombonasa to Canelos, and thence overland through the forest to Baños---a journey which occupied me exactly a hundred days. At the Indian village of Andoas, near the confluence of the Bombonasa with the Pastasa, a distant view is sometimes obtained of the Andes of Quito, but during my stay there the sky was too much obscured to allow of any but near objects being seen. On the 21st of May I reached Pacayacu, below Canelos, and was detained there three weeks in getting together Indians for conveying my goods through the forest, and procuring the necessary provisions for the way. This village stands on a plateau elevated 240 feet above the river Bombonasa, and about 1800 feet above the sea. In fine weather there is a magnificent view of the Cordillera, looking westward from the plateau, but I saw it only once for about a couple of hours in all its entirety. It takes in an angle of about 60, bounded left and right by forest on adjacent elevations. At my feet lay the valley of the Bombonasa, taking upwards a northwesterly direction; the stream itself was not visible, and audible only when swollen by rains. Beyond the Bombonasa stretched the same sort of boldly undulated plain I had remarked from Androas upwards, till reaching one long low ridge of remarkably equable height and direction (north to south): this is the watershed between the Bombonasa and Pastasa, and the latter river flows along its western foot. A little northward of west from Paca-yacu the course of the Pastasa bends abruptly, and is indicated by a deep gorge stretching westward from behind the said ridge. This gorge has on each side steep rugged hills---spurs of the Cordillera--- of from 5000 to 7000 feet high; one of those on the right is called Abítagua, and the track from Canelos to Baños passes over its summit. All this was frequently visible, but it was only when the mist rolled away from the plain, a little after sunrise, that the lofty Cordillera beyond lay in cloudless majesty. To the extreme left (south) rose Sangáy, or the volcano of Macas, remarkable for its exactly conical outline, for the snow lying on it in longitudinal stripes (apparently of no great thickness), and for the cloud of smoke continually hovering over it. A good way to the right was the loftier mountain called "El Altar" its truncated summit jagged with eight peaks of nearly equal elevation, and clad with an unbroken covering of snow, which glittered in the sun's rays like crystal---an altar to whose elevated purity no mortal offering will perhaps ever attain. [FN: El Altar seen from the western side---from Riobamba, for instance---is very distinctly perceived to be a broken-down volcano, which is by no means the case when seen from the east.] Not far to the right of El Altar, and of nearly equal altitude, stood Tunguragua, a bluff irregular peak with a rounded apex capped with snow, which also descends in streaks far down its sides. [FN: Tunguragua seen from the north and north-west is an almost symmetrical truncated cone, and the most picturesque peak in the Andes.] To the right of Tunguragua and over the summit of Abitagua appeared lofty blue ridges, here and there pointed with white, till on the extreme right was dimly visible a snowy cone of exactly the same form as Sangáy, but much more distant and loftier; this was Cotopaxi, one of the most formidable volcanoes on the face of our globe. Far behind Tunguragua, and peeping over its left shoulder, was distinctly visible a paraboloidal mass of unbroken snow; this was Chimborazo, long considered the monarch of the Andes, and though latterly certain peaks in Bolivia are said to have out topped it, it will be for ever immortalized in men's memories by its association with such names as Humboldt and La Condamine. Thus to right and left of the view I had an active volcano--- Cotopaxi I never saw clearly but once, but Sangáy was often visible when the rest of the Cordillera was veiled in clouds, and on clear nights we could distinctly see it vomiting forth flame every few minutes. The first night I passed at Paca-yacu I was startled by an explosion like that of distant cannon, and not to be mistaken for thunder; it came from Sangáy, and scarcely a day passed afterwards without my hearing the same sound once or oftener.

In the month of July 1857 I reached Baños, where I learnt that the snowy points I had observed from Paca-yacu, between Tunguragua and Cotopaxi, were the summits of a group of mountains called Llanganati, from which ran down to the Pastasa the densely-wooded ridges I saw to northward. I was further informed that these mountains abounded in all sorts of metals, and that it was universally believed the Incas had deposited an immense quantity of gold in an artificial lake on the flanks of one of the peaks at the time of the Spanish Conquest. They spoke also of one Valverde, a Spaniard, who from being poor had suddenly become very rich, which was attributed to his having married an Indian girl, whose father showed him where the treasure was hidden, and accompanied him on various occasions to bring away portions of it; and that Valverde returned to Spain, and, when on his death-bed, bequeathed the secret of his riches to the king. Many expeditions, public and private, had been made to follow the track indicated by Valverde, but no one had succeeded in reaching its terminus; and I spoke with two men in Baños who had accompanied such expeditions, and had nearly perished with cold and hunger on the páramos of Llanganati, where they had wandered for thirty days. The whole story seemed so improbable that I paid little attention to it, and I set to work to examine the vegetation of the adjacent volcano Tunguragua, at whose north-eastern foot the village of Baños is situated. In the month of September I visited Cotaló, a small village on a plateau at about two thirds of the ascent of Guayrapáta, the hill in front of Tunguragua and above the confluence of the rivers Patate and Chambo. From Cotaló, on a clear night of full moon, I saw not only Tunguragua, El Altar, Condorasto, and the Cordillera of Cubilliú, stretching southwards toward the volcano Sangáy, but also to the eastward the snowy peak of Llanganati. This is one of the few points from which Llanganati can be seen; it appears again, in a favourable state of the atmosphere, a good way up the slopes of Tunguragua and Chimborazo.

At Baños I was told also of a Spanish botanist who a great many years ago lost his life by an accident near the neighbouring town of Patate, and that several boxes belonging to him, and containing dried plants and manuscripts, had been left at Baños, where their contents were finally destroyed by insects. In the summers of the years 1858 and 1859 I visited Quito and various points in the Western Cordillera, and for many months the country was so insecure, on account of internal dissensions, that I could not leave Ambato and Riobamba, where my goods were deposited, for more than a few days together. I obtained, however, indisputable evidence that the *Derrotero* or Guide to Llanganati of Valverde had been sent by the King of Spain to the Corregidors of Tacunga and Ambato, along with a Cedula Real (Royal Warrant) commanding those functionaries to use every diligence in seeking out the treasure of the Incas. That one expedition had been headed by the Corregidor of Tacunga in person, accompanied by a friar, Padre Longo, of considerable literary reputation. The *Derrotero* was found to correspond so exactly with the actual localities, that only a person intimately acquainted with them could have drawn it up; and that it could have been fabricated by any other person who had never been out of Spain was an impossibility. This expedition had nearly reached the end of the route, when one evening the

Padre Longo disappeared mysteriously, and no traces of him could be discovered, so that whether he had fallen into a ravine near which they were encamped, or into one of the morasses which abound all over that region, is to this day unknown. After searching for the Padre in vain for some days, the expedition returned without having accomplished its object.

The Cedula Real and *Derrotero* were deposited in the archives of Tacunga, whence they disappeared about twenty years ago. So many people were admitted to copy them that at last some one, not content with a copy, carried off the originals. I have secured a copy of the *Derrotero*, bearing date August 14, 1827; but I can meet with no one who recollects the date of the original documents.

I ascertained also that the botanist above alluded to was a Don Atanasio Guzmán, who resided some time in the town of Píllaro, whence he headed many expeditions in quest of the gold of Llanganati. He made also a map of the Llanganatis, which was supposed to be still in existence. Guzmán and his companions, although they found no deposit of gold, came on the mouths of several silver and copper mines, which had been worked in the time of the Incas, and ascertained the existence of other metals and minerals. They began to work the mines at first with ardour, which soon, however, cooled down, partly in consequence of intestine quarrels, but chiefly because they became disgusted with that slow mode of acquiring wealth when there was molten gold supposed to be hidden close by; so the mines were at length all abandoned. This is said to have taken place early in the present century, but the exact date I can by no means ascertain. Guzmán is reported to have met with Humboldt, and to have shown his drawings of plants and animals to that prince of travellers. He died about 1806 or 1808, in the valley of Leytu, about four leagues eastward of Ambato, at a small farmhouse called now Leytillo, but marked on his map San Antonio. He was a somnambulist, and having one night walked out of the house while asleep, he fell clown a steep place and so perished. This is all I have been able to learn, and I fear no documents now exist which can throw any further light on the story of his life, though a botanical manuscript of his is believed to be still preserved in one of the archives of Quito. I made unceasing inquiries for the map, and at length ascertained that the actual possessor was a gentleman of Ambato, Señor Salvador Ortega, to whom I made application for it, and he had the kindness to have it brought immediately from Quito, where it was deposited, and placed in my hands; I am therefore indebted to that gentleman's kindness for the pleasure of being able to lay the accompanying copy of the map before the Geographical Society.

The original map is formed of eight small sheets of paper of rather unequal size (those of my copy exactly correspond to them), pasted on to a piece of coarse calico, the whole size being 3 feet 10 ½ inches by 2 feet 9 inches. It is very neatly painted with a fine pencil in Indian ink the roads and roofs of houses red but it has been so roughly used that it is now much dilapidated, and the names, though originally very distinctly written, are in many cases scarcely decipherable: in making them out I have availed myself of the aid of persons familiar with the localities and with the Quichua language. The attempt to combine a vertical with a horizontal projection of the natural features of the country has produced some distortion and dislocation, and though the actual outline of the mountains is intended to be represented, the heights are much exaggerated, and consequently the declivities too steep. Thus the apical angle of the cone of Cotopaxi (as I have determined it by actual measurement) is 121°, and the slope (inclination of its surface to the horizon) 29 ½° ; while on Guzmán's map the slope is 69 ¼°, so that the inclination is only three-sevenths of what he has represented it, and we may assume a corresponding correction needed in all the other mountains delineated. [FN: The apical angle of Tunguragua---the steepest mountain I ever climbed---is 92 ½° and the slope 43 ¼°]

The whole map is exceedingly minute, and the localities mostly correctly named, but there are some errors of position, both absolute and relative, such that I suppose the map to have been constructed mainly from a simple view of the country, and that no angles and very few compass-bearings have been taken. The margins of the map correspond so nearly with the actual parallels and meridians, that they may be assumed to represent the cardinal points of the compass, as on an ordinary map, without sensible error.

The country represented extends from Cotopaxi on the north to the base of Tunguragua on the south, and from the plain of Callo (at the western foot of Cotopaxi) on the west to the river Puyu, in the forest of Canelos, on the east. It includes an area of something less than an equatorial degree, namely, that comprised between 0° 40' and 1° 33' S. lat., and between 0° 10' W., and near 0° 50' E. of the meridian of Quito. In this space are represented six active volcanoes (besides Cotopaxi), viz.--

1. El Volcan de los Mulatos, east a little south from Cotopaxi, and nearly on the meridian of the Río de Ulva, which runs from Tunguragua into the Pastasa. The position of this volcano corresponds to the Quilindaña of most maps a name which does not occur on Guzmán's, nor is it known to

any of the actual residents of the country. A group of mountains running to north-east, and terminating in the volcano, is specified as the Cordillera de los Mulatos: it is separated from Cotopaxi by the Valle Vicioso.

2. El Volcan de las Margasitas, south-east by east from Los Mulatos, and a little east of north from the mouth of the Río Verde Grande. "Margasitas" (more properly Marquesitas) corresponds nearly to the term "pyrites," and is a general name for the sulphates of iron, copper, &c.

3. Zunchu-urcu, a smaller volcano than Margasitas, and at a short distance south-south-east of it. "Zunchu" is the Quichua term for mica or talc.

4. Siete-bocas, a large mountain, with seven mouths vomiting flame, south-west by south from Margasitas, west by south from Zunchu. Its southern slope is the Nevado del Atilis.

5. Gran Volcan del Topo, or Yurag-Llanganati, nearly east from Siete-bocas and south-west from Zunchu. A tall snowy peak at the head of the river Topo, and the same as I saw from Cotaló. It is the only one of the group which rises to perpetual snow, though there are many others rarely clear of snow; hence its second name Yurag (White) Llanganati. [FN:

Villavicensio gives its height as 6520 varas (17,878 English feet) in his *Geografia del Ecuador*, from a measurement (as he says) of Guzmán, but does not inform us where he obtained his information.]

[This mountain is partly shown on the extreme right margin of the map here given.]

The last four volcanoes are all near each other, and form part of what Guzmán calls the Cordillera de Yurag-urcu, or Llanganatis of the Topo.

North-east from the Volcan del Topo, and running from south-east to north-west, is the Cordillera de Yana-urcu, or the Llanganatis of the Curaray, consisting chiefly of a wooded mountain with many summits, called Rundu-uma-urcu or Sacha-Llanganati.

6. Jorobado or the Hunchback, south-southwest half west from Yurac-Llanganati, and between the river Topo and the head of the greater Río Verde.

I have conversed with people who have visited the Llanganati district as far as forty years back, and all assure me they have never seen any active volcano there; yet this by no means proves that Guzmán invented the mouths vomiting flame which appear on his map. The Abbé Velasco, writing in 1770 [FN: *Historia de Quito*], says of Tunguragua, "It is doubtful whether this mountain be a volcano or not," and yet three years afterwards it burst forth in one of the most violent eruptions ever known. I gather from the perusal of old documents that it continued to emit smoke and flame occasionally until the year 1780. Many people have assured me that smoke is still seen sometimes to issue from the crater. I was doubtful about the fact, until, having passed the night of November 10, 1857, at the height of about 8000 feet on the northern slope of the mountain, I distinctly saw at daybreak (from 5 ½ to 6 ½ A.M.) smoke issuing from the eastern edge of the truncated apex. [FN: The same morning (Nov.11), at 4 A.M., I observed a great many shooting-stars in succession, all becoming visible at the same point (about 40° from the zenith), proceeding along the arc of a circle drawn through Orions Belt and Sirius, and disappearing behind the cone of Tunguragua.] In ascending on the same side, along the course of the great stream of lava that overwhelmed the farm of Juivi and blocked up the Pastasa, below the mouth of the Patate, for eight months, we came successively on six small fumaroli, from which a stream of thin smoke is constantly issuing. People who live on the opposite side of the valley assert that they sometimes see flame hovering over these holes by night. The inhabitants of the existing farm of Juivi complain to me that they have been several times alarmed of late (especially during the months of October and November 1859) by the mountain "bramando" (roaring) at night. The volcano is plainly, therefore, only dormant, not extinct, and both Tunguragua and the Llanganatis may any day resume their activity.

Returning to the map, let us trace briefly its hydrography. The actual source of the Napo is considered to be the Río del Valle, which runs northward through the Valle Vicioso, on the eastern side of Cotopaxi. Its large tributary the Curaray (written Cunaray by Guzmán) rises only a few miles more to the south, in the Cordillera de los Mulatos, in several small streams which feed the lake Zapalá (a mile or more across) and issuing from its eastern extremity run east-south-east to Yana-cocha (Black Lake), a large body of water a league and a-half long by two miles broad. After passing this lake the river takes the name of Desaguadero de Yana-cocha, and lower down that of Río de las Sangurimas, receiving in its course (besides smaller streams) the Río de los Mulatos from the north, and a good way farther down the Río de los Llanganatis, coming from the south along a deep ravine (Cañada honda) between Rundu-umu and the Volcan del Topo. Beyond this and nearly north by east from the Volcan del Topo it is joined from the north by a considerable stream, the Curaray Segundo or Río de las Flechas, and takes the name of Río Grande de los Curarayes. The general course of the Curaray is eastward, as is also that of the Napo, and although the two rivers diverge so little from each other, they run as it were side by side through four degrees of longitude ere they meet.

The map is traversed from the north-western corner by a large stream, the Patate, rising in the western cordillera near Ilinisa, and running east-south-east through the central *callejon* (the lane between the two cordilleras) to a little south of Cotopaxi, where it reaches the base of the eastern cordillera, which it thenceforth separates from the callejon until it unites with the Chambo, at the foot of Tunguragua, to form the Pastasa. It receives all the streams flowing from the eastern side of the western cordillera, from Ilinisa to Chimborazo, of which the principal is the Ambats. From Cotopaxi the western edge of the eastern cordillera has a general direction of south by east. It consists of elevated páramos sown with lakes and morasses, and rarely covered with snow, which sink down to the river Patate, and from Píllaro southward have many deep-wooded ravines on the slope. From Píllaro northward they sink down into the plain quite bare of wood. The whole range is vulgarly called "Páramos de Píllaro." The principal tributaries of the Patate entering from these mountains are the Aláquis, which comes in a little north of Tacunya, and whose bed is subject to sudden enlargement from the melting of the snows on Cotopaxi, interrupting all communication with the capital; the Guapanti, whose sources are a number of lakes lying south of Lake Zapalá, their united waters flowing westward through the large lake Pisayambu, and entering the Patate near the village of San Miguel; and the Cusatágua, which comes down through a black wooded valley from the Cerro de los Quinuales; on the left it is joined by a stream which, about midway, forms a high cascade of two leaps, called Chorrera de Chalhuaurca (Fish-hill Fall): this cascade is visible from and nearly east of Ambato.

As the great mineral districts of Llanganati, occupying the northern half of the map, was repeatedly travelled over by Guzmán himself, it is fuller of minute detail than the rest; and I am assured by those who have visited the actual localities that not one of them is misplaced on the map; but the southern portion is much dislocated; and, as I have traversed the whole of it, I will proceed to make some remarks and corrections on this part of the map.

From Chimborazo (lying a few miles to westward of the village of Mocha) a spur or knot is sent off to the eastward, containing the mesetas or páramos of Sanancajas and Sabañán and the heights of Igualáta. Guambaló, Múlmúl, and Guayrapata, which last slopes abruptly down to the junction of the Chambo and Patate. These are so much transposed in Guzmán's map that I have omitted them in my copy, with the exception of the last. Even the environs of Ambato are much distorted; for the river Pachanlica actually unites with the Ambato a little above the mouth of the latter, instead of running direct into the Patate, some distance below the Ambato, as it is made to appear on the map.

Let us now descend the valley of the Pastasa from Guayrapata. [FN: Guayra-pata = margin (or beginning) of the wind; thus, sacha-pata = edge of the wood; Cocha-pata = margin of the lake.] The easterly wind, due to the earth's rotation, is distinctly felt along the Amazon so long as that river preserves an enormous width, and its course presents no abrupt sinuosities; but in its upper part, and on most of its tributaries, the wind is variable, and owes its modifications partly to local circumstances. In ascending the valley of the Pastasa from the roots of the Andes, one begins to feel the general wind again at a height of about 4000 feet, and, on coming out on the top of Guayrapata (9000 to 10,000 feet), the easterly wind (blowing up the gorge of Baños) strikes with tremendous force against that barrier, which is almost continually veiled in mist. The forest which crowns it is so densely hung with mosses as to be almost impenetrable; and one is forcibly struck by the contrast on emerging from the humidity and vigorous vegetation of Guayrapata to the arid sandy plains extending toward Píllaro and Ambato.

The Chambo, which flows at the base of Guayrapata, is a larger stream than the Patate (though Guzmán's map represents it much smaller), and takes its origin from the volcano Sangáy. The steep descent from Guayrapata to the river is 3000 feet in perpendicular height, and occupies the traveller two hours to descend whether mounted or on foot; but from the opposite margin of the river rises the majestic cone of Tunguragua in an unbroken slope of full 11,000 feet perpendicular! Proceeding eastwards from the confluence of the two rivers, the first stream which enters to swell their united waters is the Lligua coming from the north. Below this, and on the right bank, near the village of Baños, a small stream of tepid water, the Vascún, comes from Tunguragua. Before the last eruption of Tunguragua (April 23rd, 1773) a larger stream came down from the mountain and watered the farm of Juívi, in the angle between the Chambo and Pastasa; but the lava which descended on that side buried the farm, and since then the rivulet has been dry, though its bed is still traceable wherever not covered up by the lava. The water now finds its way through a subterranean channel, and bursts out in considerable volume on the very margin of the Pastasa, beneath the lava which is there heaped up to the height of more than a hundred feet. Not a single stream waters now the northern side of Tunguragua, all the way from Baños to Puela (half a day's journey), though several gush out of the cliff on the right bank of the Chambo.

A little above Baños, and on the same side of the river, stand a few cottages called Pitíti (the cleft), because the Pastasa at that point foams through a narrow, tortuous chasm from 150 to 200 feet deep.

Below Baños, and on the opposite (the left) bank, enters the Illúchi, whose course is parallel to that of the Lligua. The next stream, the Río de Ulva, is of considerable volume, and comes down from the snows on the north-eastern side of Tunguragua.

A very little below the mouth of the Ulva, and on the opposite bank enters a still larger stream, the Río Verde Primero, which descends from the páramos of Llanganati.

Thus far there has been little to correct in this part or the map; but the next tributary of the Pastasa therein indicated is now called the Río de Agoyán, and the farm of Agoyán occupies the site marked on the map "La Yunguílla." There is no river called Yunguílla, and the farm known by that name is actually on the farther side of the next river (the Río Blanco); while the farm of Antombós is at the eastern foot of the hill called El Sapotal, and on a smaller stream than the Río Blanco.

Exactly opposite Antombós the river Chinchin falls over a high cliff into the Pastasa.

The last bridge across the Pastasa is above the mouth of the Agoyán: on passing it we have fairly entered the Montaña, or Forest, of Canelos. A little above the mouth of the Río Blanco is the Chorrera de Agoyán, one of the finest waterfalls in South America, where the Pastasa is precipitated over a semicircular cliff, deeply excavated to the left of the fall, a height of about 150 feet.

Continuing along the left bank of the Pastasa, we next reach the Río Verde Segundo---now better known as the Río Verde Grande---which comes from the Cordillera de Pucarumi (Red-stone Ridge), running south of the snowy Llanganati. There is now a fine cane-farm near the mouth of the Río Verde, where the existing track to Canelos passes. The river is unfordable, and has to be crossed at a narrow place by throwing poles across from cliff to cliff.

The prevailing rock in the Gorge of Baños (as this deep, narrow valley may well be called) is mica-schist, though a hard, compact, black, shining, volcanic rock protrudes in many places, especially at the bridges of Baños and Agoyán.

The next river marked on the map is the Río Colorado, now known as the Río Mapóto, but well meriting its ancient name by its red margins and the red stones in its bed, coloured by a ferruginous deposit. At its mouth a broad beach (Playa de Mapóto) extends down the Pastasa for near 2 leagues: this beach is never entirely covered with water even in the highest floods; and it bears great quantities of the wax-tree called "laurél" (*Myrica cerifera*). But the Río Colorado, instead of being at the short distance from the Río Verde represented in the map, is as far apart from it as the Río Verde is from the bridge of Agoyán; and from the Río Verde to Mapóto is a good day's journey, as is also the distance from Mapóto to the river Topo. It is true that the Río Verde and Topo, though so wide apart at the mouth, may converge in the upper part (as is represented in the map); but I much suspect that the eastern portion of the map is much contracted in longitude, although, from the comparative paucity of details, the contrary might seem to be the case.

The Topo is the largest of all the upper tributaries of the Pastasa. In the time of Guzmán it seems to have been passed by a Taravita [FN: Taravita – an aerial ferry, consisting of a number of stout thongs stretched across a river from cliff to cliff, and a sort of basket slung on them, wherein a person sits to be drawn over.] a good way up, but the modern track to Canelos crosses it at only 200 yards from the mouth. The Topo, as far as anyone has been up it, is one continued rapid; and where it is crossed nothing is to be seen but rocks and foam, while the shock of its waters makes the very ground tremble. To pass over it bridges of bamboo have to be thrown from the margin to rocks in the middle, and thence to the opposite side, so that in all four bridges are needed; but a very slight flood lays one of the rocks under water, and then it is impossible to rest a bridge on it.

Only a league below the mouth of the Topo enters the Shuña, a river of little less volume than the former; but as there is a point on each side, where the rocks advance considerably into the stream, it admits of being passed by a single bridge. A flood, however, renders it equally impassable as the Topo.

When I journeyed from Canelos to Baños, I found the Shuña somewhat swollen, and crossed it with difficulty; but when I reached the Topo, I found one of the rocks, on which it is customary to rest a bridge, covered with water. My party consisted of sixteen persons, for whose sustenance every article of provision had to be carried along with us. We waited two days: the river, instead of lowering, continued to rise; our provisions were nearly exhausted, and we saw ourselves exposed to perish of hunger. In this dilemma we found a place a little higher up the river, where we determined to attempt the passage by means of three bridges. On making the experiment, we found the distance between the two rocks in the middle so great that the bamboos barely rested with their points against the side of the opposite rock instead of on the top of it; and when a man walked over them they bent with his weight into the water, whose foaming surges threatened to wash him off; and there was obviously no hope of any one passing with the load of one of my boxes. However, a thunderstorm with heavy rain came on, and, seeing no other chance of saving our lives except by risking the passage of the frail bridges, without loss of time I resolved to abandon my goods and get over to the other side. We had barely all crossed in safety when the river rose and carried away our bridges. On the third day afterwards we reached Baños, where I sought out practiced cargueros, and sent them off to the Topo; but for fifteen days from the date of my crossing it the waters did not subside sufficiently to allow of bridges being thrown over; and when the cargueros, at the end of that time, succeeded in passing to the opposite side, they found the leather covering of my boxes completely soaked and full of maggots! We had left them under ranchos of *Anthurium*-leaves (for the palms have long ago been exhausted between the Topo and the Shuña); and as the rains had been almost unceasing, the leaves had fallen off the roof upon the boxes and were rotting there. Fortunately the contents of the boxes had sustained very little injury.

Many lives have been lost in the Shuña and Topo; and of those who have fallen into the latter only one has come out alive. But the fate is more horrible of those who, shut up between the Shuña and Topo when both are so much swollen as to be impassable, perish of hunger.

The Shuña, though approaching so near the Topo at its outlet, diverges considerably in its upper part; and, as well as I can make out, its source is not far from those of the Ashpayacu and Pindu. When the Topo and Shuña are passed under favourable circumstances, the traveller on his way to Canelos arrives at an early hour the same day at the Cerro Abitagua, a steep mountain ending to the south in perpendicular cliffs, along the very base of which runs the Pastasa; so that the track is made to pass over the summit of Abitagua; and the ascent and descent on the other side occupy a whole day. The great mass of Abitagua seems alluvial; and from this point downwards no more primitive or igneous rock is seen *in situ*, nor indeed all the way down the Amazon until reaching the volcanic districts of Villa Nova and Santarem. Abitagua is also the last hill of any elevation on the eastern side of the Andes (following the valley of the Pastasa): beyond it the ground sinks in gentle undulations down to the great Amazonian plain. From its summit there is a near view of Llanganati, toward the sources of the Topo; but on two occasions that I have ascended Abitagua the summit of Llanganati has been hidden by clouds, and only its wooded flanks and deep, savage valleys have been visible. The valley of the Shuña can be traced to west and north of Abitagua. In descending the eastern slope of the mountain a fine view is obtained of the Great Plain, extending as far as the sight can reach to the south-east like a sea of emerald, in one part of which the Pastasa is seen winding like a silver band, but at so great a distance that it is impossible to discern whether its course be still obstructed by rocks and whirlpools as at the base of Abitagua.

A good day's journey beyond Abitagua brings us to the Ashpa-yacu, which is also sufficiently large to become unfordable after heavy rains: it does not appear at all on Guzmán's map. On the following day the Pindu and Púyu are reached; these rivers are equal in size to Ashpa-yacu, and the two unite at a short distance before they reach the Pastasa. In the space between them are a few huts and chacras of Jívaro Indians, the only habitations between the Río Verde and Canelos.

Beyond the Río Púyu (River of Mists) the track diverges from the Pastasa, within hearing of whose surges it has run thus far. It also passes the limits of Guzmán's map, and continues with an easterly course along the ridges which separate the basin of the Púyu from that of the Bombonasa, which latter river is finally crossed to reach the village of Canelos situated near its left bank.

Of the climate of the Forest of Canelos I can only say a few words here. The clouds heaped up against the cordillera by the wind of the earth's rotation descend in daily rains. For three or four months in the year---between November and April---the sun rather predominates over the rain, and this is called "summer;" while for the rest of the year the heavy rains allow the sun to be seen for a very brief interval each day, so they call it "winter," though the climate is in reality a perpetual spring. From the Topo eastward the mist looks as if it were permanently hung up in the trees; and beyond Abitagua wind is scarcely ever felt, except rarely an occasional hurricane; and yet after heavy rains it is customary to find the forest strewed with large green branches. Immense bunches of moss depend from the trees, hiding the very foliage; and when saturated with moisture (which no wind ever shakes out) their weight breaks off the branches whereon they are hung. I am assured by the cargueros that from this cause alone they pass through the forest with fear and trembling after heavy rains; for their load obliges them to travel in a stooping posture, so that they are unable to see the impending danger. Yet with all this moisture the climate is healthy, and I have nowhere suffered so little from going all day in wet clothes.

The track above described is one of the two routes from Ecuador to the Amazon; the other proceeds from Quito to the Indian villages on the Napo, and presents almost equal dangers and difficulties. It is easy to see that the commerce carried on by such routes must be of very slight importance. In another paper I may perhaps discuss the facilities offered and the difficulties to be overcome in the attempt to establish a safe and speedy communication between the Pacific and the Amazon by the various routes which depressions in the Cordillera seem to offer us.

I am unable to give, from personal observation, any account of the geological structure of the country represented in the central and northern portion of the map. The form of the mountains and the rugged peaks leave no doubt that trachyte is the prevailing formation; but some of the rocks seem so regularly columnar that I suppose them to be basaltic; for instance, La Mesa de Ushpa Yuras, La Capilla del Sol and El Docel de Ripalda, all near each other, and a little north of the Volcano Margasitas; El Pulpito, on the south side of the lakes at the head of the river Guapanti; El Castillejo, north-west of Sieté Bocas, &c.

The parts of the map covered with forest are represented by scattered trees, among which the following forms are easily recognizable: --

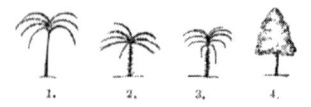

<space for tree figures 1, 2, 3, 4>

No. 1 is the Wax palm (*Palma de Ramos* of the Quitonians; *Ceroxylon andicola*, H. et B. [FN: I am doubtful if later writers are correct in referring this palm to the genus *Iriarten*.]), which I have seen on Tunguragua up to 10,000 feet. Nos. 2 and 3 are Tree-ferns (*Helechos*) the former a *Cyathea*, whose trunk (sometimes 40 feet high) is much used for uprights in houses; the latter an *Alsophila* with a prickly trunk, very frequent in the forest of Canelos about the Río Verde. No. 4 is the *Aliso* (*Betula acuminata*, Kunth), one of the most abundant trees in the Quitonian Andes; it descends on the beaches of the Pastasa to near 4000 feet, and ascends on the páramos of Tunguragua to 12,000. But there is one tree, (represented thus: ), occupying on the map a considerable range of altitude, which I cannot make out, unless it be a *Podocarpus*, of which I saw a single tree on Mount Abitagua, though a species of the same genus is abundant at the upper limit of the forest in some parts of the Western Cordillera. A large spreading tree is figured here and there in the forest of Canelos which may be the *Tocte* a true Walnut (*Juglans*), with an edible fruit rather larger than that of the European species. The remaining trees represented, especially those toward the upper limit of the forest, are mostly too much alike to admit of the supposition that any particular species was intended by them.

The abbreviations made use of in the map are : $C°$ for *Cerro* (mountain), $Cord^a$ for *Cordillera* (ridge), $Mont^a$ for *Montana* (forest), $A°$ for *Arroyo* (rivulet), $L^a$ for *Laguna*, and $C^a$ for *Cocha* (lake), $Far^n$ for *Farallón* (peak or promontory), $H^a$ for *Hacienda* (farm), and $C^1$ for *Corral* (cattle or sheep-fold).

Mule-tracks (called by the innocent natives "roads") are represented by double red lines, and footpaths by single lines. I have copied them by dotted lines.

Having now passed in review the principal physical features of the district, let us return to the *Derrotero of Valverde*, of which the following is a translation. The introductory remark, or title (not in very choice Castilian), is that of the copyist: ---

I have adhered closely to the sense and style of the original. (Guide, or Route, which Valverde left in Spain, where death overtook him, having gone from the mountains of Llanganati, which he entered many times, and carried off a great quantity of gold; and the king commanded the corregidors of Tacunga and Ambato to search for the treasure:

which order and guide are preserved in one of the offices of Tacunga.)

[]

"Placed in the town of Píllaro, ask for the farm of Moya, and sleep (the first night) a good distance above it ; and ask there for the mountain of Guapa, from whose top, if the day be fine, look to the east, so that thy back be toward the town of Ambato, and from thence thou shalt perceive the three Cerros Llanganati, in the form of a triangle, on whose declivity there is a lake, made by hand, into which the ancients threw the gold they had prepared for the ransom of the Inca when they heard of his death. From the same Cerro Guapa thou mayest see also the forest, and in it a clump of *Sangurimas* standing out of the said forest, and another clump which they call *Flechas* (arrows), and these clumps are the principal mark for the which thou shalt aim, leaving them a little on the left hand. Go forward from Guapa in the direction and with the signals indicated, and a good way ahead, having passed some cattle-farms, thou shalt come on a wide morass, over which thou must cross, and coming out on the other side thou shalt see on the left hand a short way off a *jucál* on a hill-side, through which thou must pass. Having got through the *jucál*, thou wilt see two small lakes called "Los Anteojos" (the spectacles), from having between them a point of land like to a nose.

From this place thou mayest again descry the Cerros Llanganati, the same as thou sawest them from the top of Guapa, and I warn thee to leave the said lakes on the left, and that in front of the point or "nose" there is a plain, which is the sleeping-place. There thou must leave thy horses, for they can go no farther. Following now on foot in the same direction, thou shalt come on a great black lake, the which leave on thy left hand, and beyond it seek to descend along the hill-side in such a way that thou mayest reach a ravine, down which comes a waterfall: and here thou shall find a bridge of three poles, or if it do not still exist thou shalt put another in the most convenient place and pass over it. And having gone on a little way in the forest, seek out the hut which served to sleep in or the remains of it. Having passed the night there, go on thy way the following day through the forest in the same direction, till thou reach another deep dry ravine, across which thou must throw a bridge and pass over it slowly and cautiously, for the ravine is very deep ; that is, if thou succeed not in finding the pass which exists. Go forward and look for the signs of another sleeping-place, which, I assure thee, thou canst not fail to see in the fragments of pottery and other marks, because the Indians are continually passing along there. Go on thy way, and thou shalt see a mountain which is all of *margasitas* (pyrites), the which leave on thy left hand, and I warn thee that thou must go round it in this fashion:

◠

On this side thou wilt find a *pajonál* (pasture) in a small plain, which having crossed thou wilt come on a *cañon* between two hills, which is the Way of the Inca. From thence as thou goest along thou shalt see the entrance of the *socabón* (tunnel), which is in the form of a church porch.

Having come through the cañon and gone a good distance beyond, thou wilt perceive a cascade which descends from an offshoot of the Cerro Llanganati and runs into a quaking-bog on the right hand; and without passing the stream in the said bog there is much gold, so that putting in thy hand what thou shalt gather at the bottom is grains of gold. To ascend the mountain, leave the bog and go along to the right, and pass above the cascade, going round the offshoot of the mountain. And if by chance the mouth of the socabón be closed with certain herbs which they call "Salvaje," remove them, and thou wilt find the entrance. And on the left-hand side of the mountain thou mayest see the "Guayra" (for thus the ancients called the furnace where they founded metals), which is nailed with golden nails. [FN: Query---sprinkled with gold---ED (Alfred Russel Wallace)] And to reach the third mountain, if thou canst not pass in front of the socabón, it is the same thing to pass behind it, for the water of the lake falls into it.

If thou lose thyself in the forest, seek the river, follow it on the right bank; lower down take to the beach, and thou wilt reach the canon in such sort that, although thou seek to pass it, thou wilt not find where; climb, therefore, the mountain on the right hand, and in this manner thou canst by no means miss thy way."

[]

With this document and the map before us, let us trace the attempts that have been made to reach the gold thrown away by the subjects of Atahualpa as useless when it could no longer be applied to the purpose of ransoming him from the Spaniards.

Píllaro is a somewhat smaller town than Ambato, and stands on higher ground, on the opposite side of the river Patate, at only a few miles distance, though the journey thither is much lengthened by having to pass the deep quebrada of the Patate, which occupies a full hour. The farm of Moya still exists; and the Cerro de Guapa is clearly visible to east-north-east from where I am writing. The three Llanganatis seen from the top of Guapa are supposed to be the peaks Margasitas, Zunchu, and el Volcan del Topo. The *"Sangurimas"* in the forest are described to me as trees with white foliage; but I cannot make out whether they be a species of Cecropia or of some allied genus. The *"Flechas"* are probably the gigantic arrow-cane, *Gynerium saccharoides* (*Arvoré de frecha* of the Brazilians), whose flower-stalk is the usual material for the Indian's arrows.

The morass (Cienega de Cubillin), the Jucál, [FN: *Júco* is the name of a tall, solid-stemmed grass, usually about 20 feet high, of which I have never seen flower, but I take it to be a species of *Gynerium*, differing from *G. saccharoides* in the leaves being uniformly disposed on all sides and throughout the length of the stem, whereas in *G. saccharoides* the stem is leafless below and the leaves are distichous and crowded together (almost equitant) near the apex of the stem. The *Júco* grows exclusively in the temperate and cool region, from 6000 feet upwards, and is the universal material for laths and rods in the construction of houses in the Quitonian Andes.] and the lakes called "Anteojos," with the nose of land between them, are all exactly where Valverde places them, as is also the great black lake (Yanacocha) which we must leave on the left hand. Beyond the lake we reach the waterfall (Cascada y Golpe de Limpis Pongo), of which the noise is described to me as beyond all proportion to the smallness of the volume of water. Near the waterfall a cross is set up with the remark underneath, "Muerte del Padre Longo" - this being the point from which the expedition first spoken of regressed in consequence of the Padre's sudden disappearance. Beyond this point the climate begins to be warm ; and there are parrots in the forest. The deep dry quebrada (Quebrada honda), which can be passed only at one point difficult to find, unless by throwing a bridge over it is exactly where it should be ; but beyond the mountain of Margasitas, which is shortly afterwards reached, no one has been able to proceed with certainty. The *Derrotero* directs it to be left on the left hand; but the explanatory hieroglyph puzzles everybody, as it seems to leave the mountain on the right. Accordingly, nearly all who have attempted to follow the *Derrotero* have gone to the left of Margasitas, and have failed to find any of the remaining marks signalized by Valverde. The concluding direction to those who lose their way in the forest has also been followed; and truly, after going along the right bank of the Curaray for some distance, a stream running between perpendicular cliffs (Cañada honda y Rivera de los Llanganatis) is reached, which no one has been able to cross; but though from this point the mountain to the right has been climbed, no better success has attended the adventurers.

"Socabón" is the name given in the Andes to any tunnel, natural or artificial, and also to the mouth of a mine. Perhaps the latter is meant by Valverde, though he does not direct us to enter it. The "Salvaje" which might have grown over and concealed the entrance of the Socabón is *Tillandsia usneoides*, which frequently covers trees and rocks with a beard 30 or 40 feet long.

Comparing the map with the *Derrotero*, I should conclude the cañon, "-which is the Way of the Inca," to be the upper part of the Rivera de los Llanganatis. This cañon can hardly be artificial, like the hollow way I have seen running down through the hills and woods on the western side ot the Cordillera, from the great road of Azuáy, nearly to the river Yaguachi. "Guayra" said by Valverde to be the ancient name for a smelting - furnace, is nowadays applied only to the wind. The concluding clause of this sentence, "que son tachoneados de oro," is considered by all competent persons to be a mistake for "que es tachoneado de oro."

If Margasitas be considered the first mountain of the three to which Valverde refers, then the Tembladá or Bog, out of which Valverde extracted his wealth, the Socabón and the Guayra are in the second mountain, and the lake wherein the ancients threw their gold in the third.

Difference of opinion among the gold-searchers as to the route to be pursued from Margasitas would appear also to have produced quarrels, for we find a steep hill east of that mountain, and separated from it by Mosquito Narrows (Chushpi Pongo), called by Guzmán "El Peñon de las Discordias."

If we retrace our steps from Margasitas till we reach the western margin of Yana-cocha, we find another track branching off to northward, crossing the river Zapalá at a point marked Salto de Cobos, and then following the northern shore of the lake. Then follow two steep ascents, called respectively "La Escalera" and "La Subida de Ripalda," and the track ends suddenly at the river coming from the Inca's Fountain (La Pila del Inca), with the remark, "Sublevacion de los Indios--- Salto de Guzmán," giving us to understand that the exploring party had barely crossed the river when the Indians rose against them, and that Guzmán himself repassed the river at a bound. These were probably Indians taken from the towns to carry loads and work the mines; they can hardly have been of the nation of the Curarayes, who inhabited the river somewhat lower down.

A little north and east of the Anteojos there is another route running a little farther northward and passing through the great morass of Illubamba, at the base of Los Mulatos, where we find marked El Atolladero (the Bog) de Guzmán, probably because he had slipped up to the neck in it. Beyond this the track continues north-east, and after passing the same stream as in the former route, but nearer to its source in the Inca's Fountain, there is a tambo called San Nicolas, and a cross erected near it marks the place where one of the miners met his death (Muerte de Romero). Another larger cross (La Cruz de Romero) is erected farther on at the top of a basaltic mountain called El Sotillo. At this point the track enters the Cordillera de las Margasitas, and on reaching a little to the east of the meridian of Zunchu-urcu, there is a tambo with a chapel, to which is appended the remark, "Destacamento de Ripalda y retirada per Orden Superior." Beyond the fact thus indicated, that one Ripalda had been stationed there in command of a detachment of troops, and had afterwards retired at the order of his superiors, I can give no information.

There are many mines about this station, especially those of Romero just to the north, those of Viteri to the east, and several mines of copper and silver which are not assigned to any particular owner. Not far to the east of the Destacamento is another tambo, with a cross, where I find written, "Discordia y Consonancia con Guzmán," showing that at this place Guzmán's fellow-miners quarreled with him and were afterwards reconciled. East-north-east from this, and at the same distance from it as the Destacamento, is the last tambo on this route, called El Sumadal, on the banks of a lake, near the Río de las Flechas. Beyond that river, and north of the Curaray, are the river and forests of Gancaya.

Another track, running more to the north than any of the foregoing, sets out from the village of San Miguel, and passes between Cotopaxi and Los Mulatos. Several tambos or huts for resting in are marked on the route, which ends abruptly near the Minas de Pinel (north-east from Los Mulatos), with the following remark by the author "Conspiracion contra Conrado y su accelerado regreso," so that Conrado ran away to escape from a conspiracy formed against him, but who he was, or who were his treacherous companions, it would now perhaps be impossible to ascertain.

Along these tracks travelled those who searched for mines of silver and other metals, and also for the gold thrown away by the subjects of the Inca. That the last was their principal object is rendered obvious by the carefulness with which every lake has been sounded that was at all likely to contain the supposed deposit. [The soundings of the lakes are in Spanish varas, each near 33 English inches.]

The mines of Llanganati, after having been neglected for half a century, are now being sought out again with the intention of working them; but there is no single person at the present day able to employ the labour and capital required for successfully working a silver mine, and mutual confidence is at so low an ebb in this country that companies never hold together long. Besides this, the gold of the Incas never ceases to haunt people's memories; and at this moment I am informed that a party of explorers who started from Tacunga imagine they have found the identical Green Lake of Llanganati, and are preparing to drain it dry. If we admit the truth of the tradition that the ancients smelted gold in Llanganati, it is equally certain that they extracted the precious metal in the immediate neighbourhood ; and if the Socabón of Valverde cannot at this day be discovered, it is known to every one that gold exists at a short distance, and possibly in considerable quantity, if the Ecuadoreans would only take the trouble to search for it and not leave that task to the wild Indians, who are content if, by scooping up the gravel with their hands, they can get together enough gold to fill the quill which the white man has given them as the measure of the value of the axes and lance-heads he has supplied to them on trust.

The gold region of Canelos begins on the extreme east of the map of Guzmán, in streams rising in the roots of Llanganati and flowing to the Pastasa and Curaray, [The name Curaray itself may be derived from "curi," gold.] the principal of which are the Bombonasa and Villano. These rivers and their smaller tributaries have the upper part of their course in deep ravines, furrowed in soft alluvial sandstone rock, wherein blocks and pebbles of quartz are interspersed, or interposed in distinct layers. Toward their source they are obstructed by large masses of quartz and other rocks; but as we descend the stones grow fewer, smaller, and more rounded, until toward the mouth of the Bombonasa, and thence throughout the Pastasa, not a single stone of the smallest size is to be found. The beaches of the Pastasa consist almost entirely of powdered pumice brought down from the volcano Sangáy by the river Palora. When I ascended the Bombonasa in the company of two Spaniards who had had some experience in mining, we washed for gold in the mouth of most of the rivulets that had a gravelly bottom, as also on some beaches of the river itself, and never failed to extract a few fragments of that metal. All these streams are liable to sudden and violent floods. I once saw the Bombonasa at Pucayacu, where it is not more than 40 yards wide, rise 18 feet in six hours. Every such flood brings down large masses of loose cliff, and when it subsides (which it generally does in a few hours) the Indians find a considerable quantity of gold deposited in the bed of the stream.

The gold of Canelos consists almost solely of small particles (called "chispas," sparks), but as the Indians never dig down to the base of the wet gravel, through which the larger fragments of gold necessarily percolate by their weight, it is not to be wondered at that they rarely encounter any such. Two attempts have been made, by parties of Frenchmen, to work the gold-washings of Canelos systematically. One of them failed in consequence of a quarrel which broke out among the miners themselves and resulted in the death of one of them. In the other, the river (the Lliquino) rose suddenly on them by night and carried off their canoes (in which a quantity of roughly-washed gold was heaped up), besides the Long Tom and all their other implements.

I close this memoir by an explanation of the Quichua terms which occur most frequently on the map. Spanish authors use the vowels *u* and *o* almost indiscriminately in writing Quichua names, although the latter sound does not exist in that language ; and in some words which have become grafted on the Spanish, as spoken in Peru and Ecuador, the *o* has supplanted the *u* not only in the orthography but in the actual pronunciation, as, for instance, in Pongo and Cocha, although the Indians still say "Chimbu-rasu," and not "Chimborazo" "Cutupacsi" or "Cutu-pagsi," and not "Cotopaxi." The sound of the English *w* is indicated in Spanish by *gu* or *hu*; that of the French *j* does not exist in Spanish, and is represented by *ll*, whose sound is somewhat similar; thus "Lligua" is pronounced "Jiwa." "Llanganati" is now pronounced with the Spanish sound of the *ll*, but whether this be the original mode is doubtful. An unaccented terminal *e* (as in Spanish "verde") is exceedingly rare in Indian languages, and has mostly been incorrectly used for a short *i* ; thus, if we wish to represent the exact pronunciation, we should write "Casiquiari," "Ucayáli," and "Llanganati" ---*not* Casiquiare, Ucayale, Llanganate.

"Llanganati" may come from "llanga," to touch, because the group of mountains called by that name touches on the sources of the rivers all round; thus, on Guzmán's map, we find "Llanganatis del Río Verde" "Llanganatis del Topo" "Llanganatis del Curaray," for those sections of the group which respectively touch on the Río Verde, the Topo, and the Curaray. The following are examples of the mode of using the verb "llanga." "Ama llángaichu!" – "Touch it not!" "Imapág llancángui?" -"Why do you touch it"; or "Pitag lláncaynirca?" -"Who told you to touch it?" And the answer might be "Llancanatág chári cárca llancarcáni." – "[Thinking] it might be touched, I touched it."

It is to be noted that the frequent use of the letter *g*, in place of *c*, is a provincialism of the Quitonian Andes, where (for instance) they mostly say "Inga" instead of "Inca." But in Maynas the *c* is used almost to the exclusion of the *g*; thus "yúrag," white, and "pítag," who, are pronounced respectively "yurac" and "pitac" in Maynas.

"Tungurágua" seems to come from "tungúri," the ankle-joint, which is a prominence certainly, though scarcely more like the right-angled cone of Tunguragua than the obtuse-angled cone of Cotopaxi is like a wen ("coto" or "cutu").

Of the termination "agua" (pron. "awa") I can give no explanation.

"Cungúri," in Quichua, is the knee; thus an Indian would say "Tungúri-mánta cungúli-cáma llustirishcáni urmáshpa," *i.e.* "In falling ('urmáshpa') I have scrubbed off the skin from the ankle to the knee."

Among rustics of mixed race, whose language partakes almost as much of Quichua as of Spanish, it is common to hear such expressions as "De tunguri á cunguri es una cola llaga."- "From the ankle to the knee is a continuous sore."

The following words occur repeatedly on the map:

"Ashpa" (in Maynas "Allpa"), earth. "Urcu," mountain. "Rumi," stone. "Cócha (cucha)," lake. "Yácu," river. "Ucsha," grass or grassy place ("Pajonál,"Sp.). "Póngo (pungu)," door or narrow entrance. "Cúchu," corner. "U'ma,"head. "Paccha," cataract. "Cúri," gold. "Cúlqui," silver. "Alquímia," copper. "Ushpa," ashes. "Chíri," cold. "Yúnga," warm, from which the Spaniards have formed the diminutive "Yungúilla," warmish, applied to many sites where the sugar-cane begins to flourish. "Yúrag," white. "Yána," black. "Púca," red. "Quílla," yellow.

"I'shcai," two; ex."I'shcai-guáuqui," the Two Brothers, a cloven peak to the east of Los Mulatos. "Chunga," ten; ex. "Chunga-uma," a peak with ten points, a little to south of "Ishcaiguauqui."

"Parca," double; thus a hill which seems made up of two hills united is called "Parca-urcu."

"Angas," a hawk. "Ambátu," a kind of toad. "Sácha," forest. "Cáspi," tree. "Yúras," herb. "Quínua," the "Chenopodium Quinoa," cultivated for its edible seed. "Pujín," hawthorn (various species of Crataegus) ; thus "Montaña de Pujines," Hawthorn Forest; "Cerro Pujin el chico," Little Hawthorn-hill. "Cubiliín," a sort of Lupine, found only on the highest páramos. It gives its name to a long ridge of the Eastern. Cordillera, mostly covered with snow, extending from Condorasto and El Altar toward Sangay. "Totorra," a large bulrush from which mats are made; hence "Totorrál," amarsh full of bulrushes. "Sara,"[1] maize.

"Tópo" is the name given in Maynas to the Raft-wood trees, species of Ochroma (of the N.O. Bombaceae). They begin to be found as soon as we reach a hot climate, say from 3000 feet elevation downwards.

"Rundu," sleet; thus "Rundu-uma" Sleety Head. "Rásu" is snow, and occurs in "Chimbu-rasu," "Caraguai-rasu" (Carguairago), and many other names. The vulgar name for snow as it falls is "Papa-cara," *i.e.* potato peelings.

"Pucará" indicates the site of a hill-fort of the Incas, of which a great many are scattered through the Quitonian Andes.

[]

## *Travels in Ecuador* Frontispiece

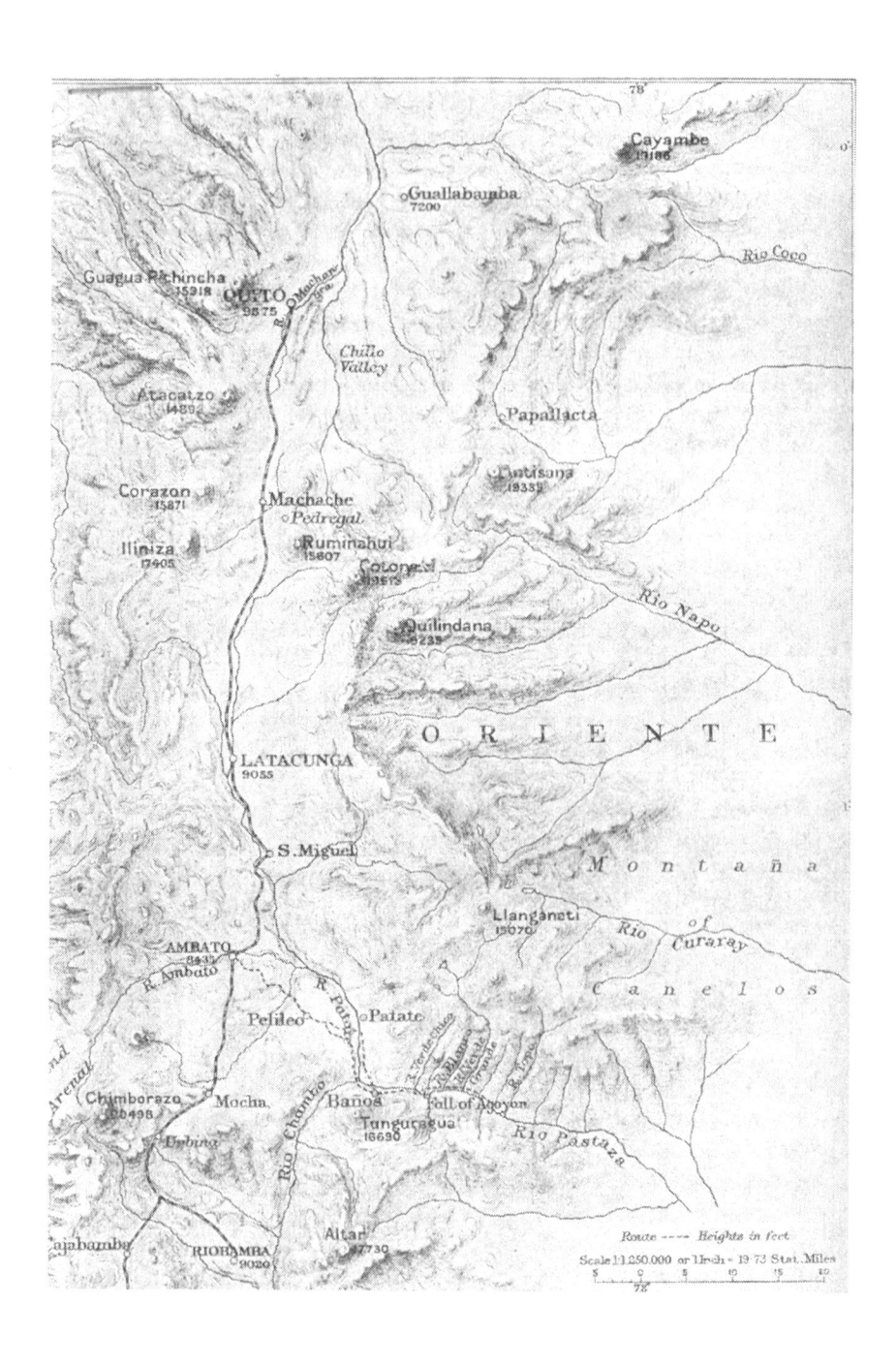

# Travels in Ecuador

### By: JORDAN HERBERT STABLER

Paper originally published in:
*Journal of the Royal Geographical Society 1917*

AT the time of the discussion in the Spanish Council of State concerning the decision of the Ecuadorian-Peruvian Boundary dispute, upon which His Majesty the King of Spain had been asked to arbitrate, the President of the Council is said to have remarked, "Beyond the natural boundaries, raised by the hand of God, it will be difficult for Ecuador to hold territory, for she is the Switzerland of the Americas." This remark is indeed true, for a striking similarity exists between the Alpine Republic and this country in the Andes — both having as common features lofty snow-clad peaks, deep ravines, and broad valleys enclosed by great mountain ranges — and the comparison enables one to form a topographical idea of Ecuador.

The Republic of Ecuador, which with the exception of Uruguay is the smallest of all the South American Republics, lies between 2° N. and 6° S. lat. The longitudinal extension of the Republic is as yet undecided, for, although many years of discussion have passed and many decisions have been made, the Eastern and Southern frontier limits are still in dispute with the neighbouring Republic of Peru, and some territory is in litigation with Colombia.

One is almost safe in claiming that Ecuador has more boundaries than any other country, for there are maps of the Republic showing six different frontiers according to six different opinions. There are the limits claimed by the Government of the Republic, which take in the greatest extent of territory and stretch far to the east, including a vast portion of the "Oriente," that territory lying to the east in the great Amazonian plain. This delineation of the frontier was made by Restrepo and Humboldt in the eighteenth century. Another frontier line is that known as the Pedemonte Mosquera line, and was drawn in 1830. The third is the provisional boundary made according to Menendez-Pidal, the Spanish High Commissioner, in 1887. The fourth is the line drawn according to the ideas set forth in the Garcia Herrera agreement. The fifth is the boundary as outlined in 1909 by the Spanish Council of State. The sixth is that claimed by the Government of Peru. These limits of Ecuador are embodied in a chart issued by order of the President of the Republic. When rumours were circulated as to what the decision of the King of Spain would be, both countries made objections; the King resigned his position as arbitrator, and the boundaries still remain in *statu quo.*

The Republic is divided into three distinct divisions from west to east, clearly defined by the great Cordilleras of the Andes. They are the Pacific littoral sloping up to the Western Range some 60 to 80 miles; the great Inter-Andine plateau, at an altitude of from 7250 to 9200 feet, in some places over 100 miles broad, fertile, cultivated, good grazing country; and, thirdly, the country known as the "Oriente" stretching from the Eastern Cordillera to the farthest border of Ecuador, tropical jungle country, unexplored to a great extent and known only to the semi-savage Indian tribes, to a few travellers, to the Ecuadorian officials at government posts, and to the "Caucheros," rubber hunters, who make a yearly trip to the interior.

From Guayaquil; the principal port of Ecuador, on the Guayas River, 2° South, the journey to Quito, the capital, is now made in two days by the trans-Andine railroad, a much easier trip than in Whymper's time in 1880, when it necessitated from six to fifteen days by mule according to season, over almost impassable trails, with no accommodation, and with but little or no food to be found on the way. Nevertheless the trip is still full of interest and of the unexpected. One must be provided with an abundant supply of food and warm blankets, for landslides and derailments are frequent, and a night spent on an Andine Pass without warm covering and nourishment is none too pleasant.

From the banks of the Guayas River the railway runs inland some 60 miles through coco, banana, and tagua plantations, and through thick tropical jungles, abounding in palms of all descriptions; and then begins, to ascend the Andine slopes through the valley of the Chan-chan River, which has its source in the lower hills of Chimborazo. Reaching the outer walls of the Cordillera it quickly mounts to the high plains by means of a "switchback" track cut into the side of an almost perpendicular cliff by a skillful feat of engineering.

The western wall of the Andes once surmounted, the track runs north, crossing at an altitude of 11,362 feet the sandy wind-swept plain known as the "Grand Arenal," which is, at almost all seasons of the year, a prey to the snow and wind storms that come with deadly blasts from the high slopes of Chimborazo. Here, in the days before the railway, many travellers were frozen to death in the severity of these storms, and it present the trail is dotted with the bones of pack-animals.

Leaving the sandy plateau and winding through the valleys of the outlying slopes of the mighty Chimborazo, the railway at length comes out upon a broad plain, at the end of which is the capital of the Province of Chimborazo, Riobamba by name, a quaint old colonial Spanish town with streets exceptionally wide as a precaution against earthquake, built on sandy soil, at an altitude of some 9030 feet. Here the night is passed, for travelling by rail at night in Ecuador is not considered safe. The volcano of Altar, rising to 17,730 feet according to the observations of Reiss and Stübel, lies almost due east of the town and surmounts this part of the eastern Cordillera.

From Riobamba the railway passes due north along the Inter-Andine plain, leaving the great mass of the ranges of Chimborazo to the west until it reaches the town of Ambato. From this point the road-bed descends a little until the town of Latacunga is reached, and passing over the Páramos of Cotopaxi, winds up through the ridges of the eastern slopes of the Andes, where an excellent view is obtained of the great peaks of Iliniza, Corazon, Antisana, Rumiñahui, and Atacatzo. The run into Quito from there on is down grade, and one arrives at the small wooden station on the outskirts of the town, barring landslides and derailments, in the late afternoon of the second day of the journey.

Quito, the capital of Ecuador, is at an altitude of 9342 feet, according to the observations of Whymper, while the survey of the railway engineer makes it some 250 feet higher. It lies close south of the equator at a distance of about 15 English miles. It is beyond doubt one of the most interesting and picturesque cities in the western hemisphere, for it still retains the charm of colonial days, and the modernizing influence of the outside world has as yet touched it but lightly. The northern capital of the Inca Empire, captured by the Conquistadores after their almost unbelievable marches over the Andes, the seat of the Vice-Regal Governor of the Presidency of Quito, the scene of some of the earlier of the attempts for independence, and, after the formation of the Republic, the theater of much revolutionary activity, Quito has a history of great importance, in the development of Spanish America.

The many plazas; the monasteries of the Dominicans, the Mercedarios, the Franciscans, and other of the great Orders; the great patios of the houses of the descendants of noble Spanish families; the religious processions frequently passing through the streets; the variegated colour scheme formed by the bright ponchos of the Indians of the city and the orange-coloured Macanas of the tribes of the hills and north country, imprint an indelible picture upon the mind. Looking down upon the city from the slopes of the volcano of Pichincha---the mountain which dominates the town---one sees below a wide extent of closely joined roofs, with here and there the tower of some great church or monastery; for Quito is for its size one of the strongest Catholic cities in South America, having some two hundred churches, chapels and monasteries. The city covers a wide area; but it is very difficult to form an idea of its population, as is so often the case in Spanish-American cities where no regular census can be taken. Although a population of some eighty thousand is claimed, I should consider a conservative estimate was from forty to fifty-five thousand.

The population may be divided into three distinct classes: the pure Indian, the descendant of the Quichua tribes speaking that language and a little Spanish, who are in the majority; the "Cholos," or mixed class---Indian and Spanish; and lastly, the pure Spanish families, who have come down in direct line from the Conquistadores. The principal streets show very well the general character of the city: the two-storied houses and the flag-paved streets, under almost all of which run streams of water from the sides of Pichincha, draining the city through the deep volcanic ravines or "quebradas."

The hill known as the "panecillo" or "little loaf" at the south-west end of the town, rises some 300 feet in height. It is reputed locally to have been built by order of the Inca as a tomb of one of the kings. The peaks of Cayambe, Imbabura, Cotocachi, all to the north, and Atacatzo, Corazon, Antisana, and Cotopaxi to the south, are visible from this point on a clear morning of the dry season.

Quito has one of the most regular climates of any capital in the world, and this has been proved by the observations made at the observatory erected by the French mission in the park of the city. The mean annual temperature is 58.8° Fahr., the maximum annual is 70°, and the minimum annual is 45°. The average range in the twenty-four hours is some 10°.

During the two years I spent there I found that I never had to worry about what the weather was going to be. One rarely made a mistake, as the weather conditions seemed to change as if by clockwork. In the summer months, from October to April - - - the rainy season - - - the rain commences to fall in a torrential downpour regularly at a little after two p.m., and by five or six it has usually cleared off and the nights are almost always cloudless. From May until the latter part of September it is clear and very dry, and quite cold in the early morning and late evening. I have known an occasional shower and once or twice a hailstorm in the winter months. There were very few rainy mornings, even in the wet season, during all the time I was in the highlands of Ecuador, and I noted very few days when it rained all day even in the middle of the wet season. Hailstorms are fairly frequent, but only last for a quarter to half an hour.

My travels from Quito into the little-known parts of the Republic were almost always made with Dr. Pierre Reimbourg, a Frenchman who has spent some years in Ecuador and has made travels from Quito into the little-known parts of the Republic were almost always made with Dr. Pierre Reimbourg, a Frenchman who has spent some years in Ecuador and has made observations for the Ministère de l'Instruction Publique, and with M. Paul Suzor, the Secretary of the French Legation. These companions of many excellent and interesting expeditions are both serving their country at the Front, and I have no doubt that they are as hardy and unflinching in the supreme test as they were in former moments of minor difficulties on the Andine trails.

One of the most interesting trips which may be made from Quito, in a very short time and with but little hardship, is the ascent of the now extinct volcano of Pichincha, the summit of which is at an altitude of 15,918 feet. The ascent may be made to one of the lower peaks almost all the way on horseback, and if one goes the night before to a hacienda some three hours from the city one may sleep there and go up to the summit and back in a day.

On the ascent four distinct belts of vegetation may be observed: (1) The lower slopes---with some few myrtles and Eucalyptus trees, and fields under cultivation with wheat, barley, and potatoes. (2) Shrubs of many varieties. (3) The Páramo---between 12,000 and 14,000 feet---pasture country. (4) Grass; this is intermingled with some hardy shrubs distributed in scattered patches. The plant known as *Lupinus alopecuriodes* is characteristic of this region.

As one climbs over the outlying slopes a superb view of the Andine plain is obtained. A large waterfall is passed far up the side of the mountain, and, one reaches quite soon a height overlooking a sea of clouds.

The expedition necessitates much more time if one desires to make a descent into the crater of the now extinct volcano. Indian guides must be procured, and a camp made just below the summit of the Guagua Pichincha, one of the two peaks of the mountain. According to the observation of Dr. Reimbourg the diameter of the crater is about 1500 feet, but it was impossible to obtain an exact measurement, as the clouds prevented observations to a great extent. The greatest depth of the crater, according to Professor W. Jameson, who visited Pichincha, is 2460 feet. One may descend by means of ropes to a floor some 500 feet in depth where there are traces of sulphur and some small shrubs.

A two days' journey to the north-east brings one to the great mountain of Cayambe on the equator. For several miles the route follows the Camino del Norte, which runs from Quito to the Colombian frontier then on to Santa Fé de Bogata, a journey of some thirty-five days on horseback. It is always full of interest and typical of the life in the high Andes. Indians in bright costumes run along at their regular trot; women with babies on their backs, and men bending under their loads, which are held by a broad strap over their forehead, are continually passing or are to be seen drinking "chicha" as the national beer is called, at the little posadas on the side of the road. Leaving the Camino del Norte the road to Cayambe runs to the east and crosses the great "quebrado" of Guallabamba, 7200 feet, which Whymper considered to be the biggest earthquake fissure in Equatorial America. This ravine is infamous for its fevers, and many prayers are said before any native crosses it in the rainy season. In this valley are grown sugar-cane, chiromoya, lemons, and other fruit of the temperate zone.

Cayambe is a wonderful mass, rising to an altitude of 19,186 feet in eastern Cordillera. It is so immense that one easily imagines that it covers the greater part of the northern half of the Republic. Its lower slopes are considered among the best pastures in the highlands, and great herds of cattle and wiry Ecuadorian horses graze here. Some Ecuadorians say that there are over 40,000 head of cattle on the haciendas on the slopes of the mountain.

The smaller variety of the Ecuadorian deer are to be found in the páramos of Cayambe (9000 to 11,000 feet), and are tracked with the aid of the big hounds bred on this mountain. The Indians of this region are good hunters, and one is surprised at their hardiness and strength. The typical costume is a pair of cotton trousers and a cotton shirt, and one or two ponchos of varying thicknesses. They all wear the native sandal "alpargata" and a wide felt hat. But in spite of the thinness of their covering they never seem to feel the extreme cold of the páramos.

The expedition to the mountain of Iliniza, a large mountain in the western chain to the south-west of Quito, is one of the most interesting from the point of view of sport. The route out of Quito leads along the Camino Real, the Vice-Regal "royal road" running to Riobamba. This road was built in the early days of the Conquest, and crosses in several places parts of the Inca road from Cuzco to Quito. It was remade in 1872 by Garcia Moreno, then President of Ecuador.

Arriving at Machache, a small town close to the railroad, one takes a trail leading to the right and ascends from the valley to the western slopes. The peak known to Whymper as the "Little Iliniza" rises to an altitude of 16,936 feet, and is one of the most prominent of the western Cordillera. On its lower slopes deer and wood-pigeons and sometimes partridges are to be found.

To the south-east of the capital, between the hamlet and hacienda of Pedregal (11,629 feet) and the mountain of Antisana, one of the greatest peaks of the eastern range (19,335 feet), lies some of the best shooting country in the Andes. Besides the Andine deer are found tapir, known in Ecuador as danta, and a lake affords wild duck. The páramos in this region are very exposed, there is much rain and snow, and the journey is a difficult one. The deer which are found on the páramos of the Ecuadorian Andes are the *Odocoileus peruvianus*. This is an ally of the Virginian deer. The Andine tapir is the *Tapirus pinchaque*, allied to the Amazonian tapir, but with a thicker coat. The turkey is probably the *Meleagris gallopavo*.

In the month of September, 1910, M. Suzor, Dr. Reimbourg and I set out from Quito to make a trip to the town of Baños, the southern gateway to the Oriente on the Pastaza River, and lying under the slopes of the volcano of Tunguragua. Dr. Reimbourg wished to make certain observations on the volcano for the Ministère de l'Instruction Publique, and I had the intention of continuing further on down the Pastaza River into the Montaña of Canelos, as this part of the Oriente is called.

We left Quito in the early morning by train and arrived at the town of Ambato, where we were to procure horses and pack-animals for our trip. At about 10 o'clock, passing close by the active volcano of Cotopaxi, we were afforded a somewhat rare view of the cone, the summit of the mountain being free from clouds. Great volumes of black smoke were pouring out, and a portion of the side of the mountain was jet black where the ashes and lava had melted the snow---a striking contrast to the other sides, which were dazzling white in the sunlight. Cotopaxi is a most satisfactory volcano in that one is rarely disappointed in seeing it in eruption; and from the higher parts of Quito one can see almost every day in the sky a long line of black made from its smoke and by night a red glare on the horizon.

Ambato is a pleasant town some 45 miles south of Quito. It lies in the Inter-Andine plain at an altitude of about 8435 feet on a sandy plateau. In the dry season the town is wind-swept and dusty. There are few trees in the town itself, but the outskirts have been irrigated, and there are orchards and gardens and trees along the banks of the Río Ambato, a small stream which runs through a narrow valley close to the town. Ambato is considered by the Ecuadorian's to be by far the prettiest city in the Republic, and is famous for its fruits, which are excellent and abundant. Oranges, chirimoyas, aguacates, granadillas, are grown here, as well as pears and peaches and other fruits of a more temperate zone.

The people of Ambato lay claim to a population of ten thousand, but the most careful estimates I have seen do not concede it over six thousand inhabitants. The climate is healthy, and numbers of people come to Ambato from Quito and from the coast for a change of air.

Notwithstanding the claims which the people of Ambato make as to the longevity of its citizens and the healthfulness of the place, when rumours spread about that a foreign physician was in the town, Dr. Reimbourg had many calls made upon him for medical advice, and his kindness was rewarded by gifts of "dulce de guyava," a sweet made from the guyava, for which Ambato is famous. These sweets formed a valuable addition to our stock of provisions later on in our journey, as we found that tough beef and a watery potato-soup called "locro" constituted the principal diet of the people of Baños.

We left Ambato at 6 a.m. by the road leading to the east. Winding up out of the valley of the Río Ambato one obtained an excellent view of the mountains of Chimborazo and Tunguragua, and looking down upon the town all that was seen was a green spot in the waste.

The trail to Pelileo runs south-west from Ambato through almost desert country. Both sides of the sandy road are lined with cactus plants and a species of American aloe. The air is exceedingly dry and the glare of the sunlight so strong that I found a pair of smoked spectacles and a sun-helmet indispensable. Ecuadorians cover their faces with veils or with large handkerchiefs when travelling through this country, as they consider sunburn dangerous.

Pelileo, which we reached at midday, is a small town with some 1500 to 2000 inhabitants as far as we could ascertain; and almost every person we saw was either pure Indian or had Indian blood. Its principal houses and churches are built of a volcanic rock and grey pumice stone. It is an old Spanish town and some of the churches are good examples of colonial architecture.

On the banks of a small stream, which, flowing from a spring in the ridges above Pelileo runs into the Río Patate, we stopped for lunch and lifted our cups and drank to the Amazonas, for were we not practically at its very headwaters?

The Patate flows into the Pastaza, the Pastaza into the Marañon, and the Marañon further along its course becomes one with the Amazon itself. Following the course of this stream we came to the Río Patate, which we crossed and continued along its valley which is beautiful and enjoys a delightful climate, being some 3500 feet below Quito and sheltered from the cold winds by the ridges of Tunguagua, which with its snow-capped peak towers far above this region. In this valley coffee and sugar-cane are grown in abundance and one of the wealthy families of Quito owns several haciendas along the banks of the river. Not far from the junction of the Río Patate and the Río Chambo, which unite to form the great Río Pastaza, the trail ascends from the valley and follows the contour of the slopes above the "Puente del Union," as the bridge at the meeting-place of the two rivers is called. It then leads along the hills above the Pastaza, being in some places almost impassable, and further on winds down to the bank of the river. The Pastaza is crossed a mile and a half from Baños by means of a small bridge across the gorge 300 feet deep, which it has cut through the solid rock, and where it rushes through the narrow channel churning up white foam.

Baños, which derives its name from the hot baths which are located close to the town under the foothills of Tunguragua, is a small village of some nine hundred to one thousand souls, lying close under the volcano, and according to our aneroid at an altitude of 8750 feet. The inhabitants are mostly of the Cholo type, but on market days the town is crowded with Indians from the hill villages. Jivaro Indians from the Oriente, usually in parties of three or four, often come into the town to trade. Their settlements are along the Pastaza River and its tributaries. They are semi-savage but quite friendly to white men, and those who come as far west as Baños have picked up a few words of Spanish. They are stocky well-built men, with rather round faces and blunt features of a dark bronze colour. Their faces are usually painted with a few marks of some sort of black paint, and they wear their straight black hair cut just below the base of their neck. They wear little or no clothing in the montaña, but on reaching a town dress themselves in what looks like a football jersey and cotton trousers. They carry long poles of bamboo, and some bring with them their blowpipes---six or seven feet in length---through which they blow clay pellets or darts. They always carry on their backs a finely woven basket which contains their food for the journey.

The houses of Baños are made of bamboo and mud, with palm-leaf roofs, and everything is of most primitive nature. There is a small church and a little four-roomed monastery in charge of two Belgian priests of the Dominican order, one of whom, Padre Van Schoote, formerly an officer in the Belgian army, had spent nineteen years as a preaching Father among the Indians of the Oriente. He had passed eight years at Maccas, which lies far to the east of Ríobamba in the heart of the slopes of the Eastern Andes, among the head-hunting tribes, and eleven years among the Jivaros at Cañelos in the Oriente.

A small hospicio or rest-house, a well-built stone building with several small bare rooms, is attached to the monastery, and here we made ourselves comfortable during our stay at Baños. We were very content to install ourselves in such a shelter; as this region is exceedingly damp and there is a great deal of rain. It is never very warm, and in the early morning the temperature is between 40 and 50 Fahr., and the thermometer does not even rise above 70 at midday.

The waters which come out at a high temperature from the springs in the hills under the volcano are said to be a cure for rheumatism and gout. The baths are prepared in a primitive manner. An Indian is sent to dig a deep hole close to the streams of hot water, and by making a little canal for the hot water and a similar one for the water from a cold stream close by, the bath is filled and regulated.

The patron saint of Baños, known as "La Virgen de las Aguas Santas," has a great reputation all over Ecuador as a worker of miracles, and on her festival the town is crowded with Indians who come many days' journey from the northern and southern parts of the Republic.

The eastern Cordillera of the Andes, which runs approximately north and south throughout the length of Ecuador, forms an almost impenetrable barrier between the Inter-Andine Plain and the Oriente. Through this barrier there are but few passes, and the three most accessible are those by way of the town of Papallacta and the Río Napo, east of Quito, by way of Cuenca and the Río Paute to the far south of the Republic, and via Baños and the Río Pastaza. This last is probably the most practicable gateway to the Oriente. Through these passes the Ecuadorian Government must send its officials and soldiers and provisions, and the task is a very difficult one. To get to the centre of the Oriente takes from three to six weeks, and ten to fifteen days at least must be made on foot over trails so deep in mud that it takes many hours to go a few miles. The question of transportation is also a large factor in any expedition into this country. An Indian bearer can only carry from twenty-five to forty pounds over these trails, and all provisions must be carried into the Oriente; but bearers are not plentiful and are most unsatisfactory. On the other hand, the Peruvians have a relatively easy access to the Amazon Plain, for their launches run from Iquitos up and down the rivers flowing into the Marañon, and men and supplies may be moved with facility. It is said that there are now many Peruvians in the part of the Oriente claimed by Ecuador who have settled on the smaller rivers, and who are armed to resist ejection by the Ecuadorian officials. What the final boundary of the Republic will be it is hard to say, but Peru has lost no time in endeavouring to send as many of her citizens as she is able into the disputed territory.

A railway from Ambato along the valley of the Pastaza to the Curaray River has been projected and a survey made of certain parts. The road construction was commenced at Ambato under the direction of two engineers, Messrs. Moore and Fox, in 1912, and I understand it has been graded for a few miles to the east. It will probably be a great many years before such railway is completed, for this route presents almost unsurmountable difficulties.

From Baños I set out to the east over the Oriente trail which leads to the village of Canelos on the Río Bombonasa. The road, which is passable for mules as far as the waterfall of

Agoyan, keeps close to the banks of the Pastaza after leaving the town. There are three bridges over the river just below Baños. These are some 250 feet above the river, and are built by means of great logs pushed out from each side, and another log or two logs spliced together between. It is rather ticklish work crossing them, especially if there is a strong wind in the gorge, as often happens. The road leads close by the river, through sugar-cane plantations, with here and there a "trapiche" or cane mill, by the side of groves of plantains and palm trees and by patches of *camote*, as a vegetable of the potato family is locally known.

Some miles further on the Pastaza is crossed by a well-built bridge constructed by Padre van Schoote, and the waterfall of Agoyan is reached. This point in the trail is at an altitude of about 5500 feet. This waterfall is the largest in the Oriente, and as the river has cut a deep channel into the solid rock and comes down with great force, it is a beautiful sight from the trail. This waterfall marks the beginning of the Montaña of Canelos, the entrance into the real Oriente.

The Montaña of Canelos, the forest on the edge of the Amazon plain which Richard Spruce, according to Mr. Wallace's *Notes of a Botanist*, claimed was "the most cryptogamic locality on the surface of the globe," is bounded on the west by the volcanoes of Cotopaxi, Llanganati, and Tunguragua, and on the east by the slopes of the Amazonian lowlands. Through this forest Gonzalo Pizarro wandered nearly two years in search of cities "as rich in gold as those of all Peru," and returned with only 80 members out of a company of Spaniards and Indians numbering 4500. Ferns, mosses, and lichen grow in the forest in great profusion. Of the ferns the Genera *Marattia* and of the mosses the Genera *Hookeria* were most abundant.

After leaving Agoyán the trail becomes a track 3 feet wide, very rough and with deep mud holes, and the progress is slow. The undergrowth is very thick, all or the jungle is moist, and it rains at frequent intervals. There is a light mist continually overhead. The palm trees are numerous all through the region. Spruce found that the *Iriartea ventricosa* was the most abundant species. There are also some wax palms, the *Iriartea andicola*. There are many plantains, and the undergrowth is very thick. In the season I was there I noticed very few orchids. Several small rivers are crossed on the way from Baños to the Río Verde, notably the Río Blanco and the Río Verde Chico; but all may be forded.

At 15 English miles from Baños the Río Verde Grande joins the Pastaza and near its bank there is a "trapiche" for grinding sugar-cane and making aguardiente. This is the last building to the east in the Montaña of Canelos with any pretensions to civilized architecture. These 15 miles from Baños to the Río Verde are the longest I have ever travelled, for the mud of the narrow trail is so deep and sticky that to go a mile sometimes takes over an hour. In the bad rains the trip can hardly be made under two days, and this trail is typical of all trails in the Oriente. One must travel as lightly equipped as possible.

The Río Verde, as its name implies, is of a deep green colour, and flows due south from the Llanganati Mountains along a steep valley, the course of which has yet to be explored. The junction of this river with the Río Pastaza is remarkable, for it comes with great force down a hanging valley whose sill is some 60 feet above the bed of the Pastaza. The momentum of the water carries it across, forming an arc, which cuts the far bank of the Pastaza where it has eroded a bay, whence the water of the Verde is turned into the Pastaza.

Spruce states in his notes, made in 1857, that the cascade formed by the Río Verde at this point is some 200 feet in height. I think that this is an overestimate, and that the erosion of the river valley cannot have progressed as rapidly as would be implied.

The Río Verde is locally reputed to have its source in the lake in the Llanganati Mountains, at the bottom of which the golden vessels which formed the ransom of King Atahualpa were thrown by the Incas when news reached Quito that their ruler had been murdered.

In the paper by Richard Spruce read before this Society in 1860, and published in the *Proceedings R. G. S.*, 1861, p. 163, the story of the Inca treasure in the Llanganati is told at length. A short résumé of the story is as follows: The contents of "the Chamber filled with Gold" stored in Quito for the ransom of the king from the Spaniards was carried swiftly into the Eastern Andes by Inca runners when the messengers announced the murder of the ruler. A river which flows through a valley among these mountains was dammed and the gold thrown into the artificial lake so formed. A Spaniard Valverde by name who many years afterward married an Indian woman, or as some say an Indian princess, was given the secret of this treasure by his wife. He made many trips to the lake and must have been successful in his search, for he returned to Spain a very wealthy man and bequeathed the secret of the lake to his king upon his death. The key to the treasure, or the *Derrotero of Valverde* as the guide is called, was sent by the King of Spain to officials at the town of Latacunga with instructions to make a search. Many attempts were made in colonial days as well as later; but as yet it appears that no one has found the hidden lake. During my stay in Ecuador, and some months prior to my trip to the Montaña of Canelos, a compatriot of mine . . . Major Brooks by name . . . was able to procure a copy of the *Derrotero* at Latacunga, and from it constructed a map with the purpose of searching for the lake.

He made two attempts, starting from the town of Ambato; but on the first attempt his Indian carriers deserted him in the heart of the mountains, and he was only able after great difficulty and with the aid of his personal servant to return to his starting-point. His second attempt was not more successful, for while he was able to find his way to certain points in the Llanganati range marked in the guide, and reached a lake, his camp was flooded and most of his provisions ruined or lost by a sudden rising of the lake, and he had to give up the search. On his return he told me that he was not sure whether he had reached the lake containing the treasure, for he himself, as well as all the other treasure-seekers who had followed Valverde's guide, had been mystified by certain directions.

These are, to quote from Spruce's translation, "Go forward and look for the signs of another sleeping-place, which I assure thee thou canst not fail to see in the fragments of pottery and other marks, because the Indians are continually passing along there. Go on thy way, and thou shalt see a mountain which is all of margasitas (pyrites), the which leave on thy left hand, and I warn thee that thou must go round it in this fashion: (*Diagram showing clockwise direction of motion, inconsistent with what precedes.*) On this side thou wilt find a pajonal (pasture) in a small plain, which having crossed thou wilt come on a cañon between two hills, which is the Way of the Inca." The question of turning to the left or right of the mountain appears as yet to be unanswered.

I hoped to be able to push up along the bank of the Río Verde in the direction of the Llanganati Range, but this project had to be given up as I was called back to Baños by a messenger sent by the good padres, to find that I must return at once to Quito.

To the geographer, the traveller, and explorer Ecuador presents a great range of interest. Most of the country has not been mapped, a great deal is still unexplored, and a vast amount of valuable work remains to be done. In concluding this paper I wish to add that I most heartily recommend the "Switzerland of the Americas" as a field for geographical investigations the result of which I feel sure will be of lasting service to the science.

[]

# The Inca Treasure of Llanganati

By: EDWARDS CRANSTON BROOKS

Correspondence published in:
*Journal of the Royal Geographical Society 1918*

Stabler's paper drew an almost immediate response from an American... E. C. Brooks. The Royal Geographic Society quickly published Mr. Brooks correspondence in its January 1918 *Geographical Journal*, volume 51, No.1, page 59, entitled *The Inca Treasure of Llanganati*, reproduced below in its entirety...

"I beg to bring to the attention of your society the enclosed copy of an editorial in the *New York Sun* for November 4 and my reply thereto in the issue of the 14th, which I also enclose.

Since it was in your journal that Mr. Spruce's article and Señor Guzmán's map were first published, I have always intended to send you a report of my discoveries; but I hoped to make a third journey; and, indeed, for the last three years plans have been made each autumn for my going. The financial man has, however, each time been prevented from going, although last year he bought the outfit recommended by me, and now has it ready for next year.

Mr. Stabler's article has taken me by surprise and has rather forced my hand before I was ready to go into print. However, I think well, in case anything happened to me, to have it go on record that the Valverde *Derrotero* as published first in your *Journal* (about March 1860) is substantially correct and surprisingly so considering that it was written on a deathbed in Spain and after some years of absence of the writer from Ecuador."

[The article in the *New York Sun* of November 4 is based on the passages in Mr. Stabler's paper (*G. J.*, October 1917, 50, 251) which summarize the story of the Inca treasure, and mention the unsuccessful search made by Major Brooks, whose letter to the *Sun* of November 12 we have the pleasure to print below. Major Brooks has kindly undertaken to give us further news of his quest at a later date.---Ed. G. J.]

[*New York Sun*, 12 Nov. 1917.]

## THE WAY TO THE GOLD OF THE INCA
## LIES OPEN TO ALL

*A Hunter of the Treasure which was to ransom Atahualpa*

*tells of his Adventures in Ecuadorean Mountains.*

More than anyone else, perhaps, I was interested. in the editorial article in *The Sun* of Sunday, 4 November 1917, entitled *The Way to the Gold of the Inca*, because I am the Major Brooks spoken of, who made two expeditions into that part of the Ecuadorean Andes called the Llanganati Mountains, in which there is much reason to believe the hidden treasure of the Incas is located.

I wish to correct some errors which seem to be in the original article which you discuss. I make these corrections with the most kindly feelings toward Mr. Stabler, the author, whose friendship. I esteem most highly. His memory is a little at fault as to some parts of what I related to him, but he cannot be blamed for this, considering the time that has elapsed since 1910, when I made my trips into that wilderness.

It must be conceded that the one who underwent the peculiar hardships of climbing in those rough and uninhabitable regions must have had the incidents more indelibly impressed upon his memory than they could have been impressed upon the memory of anyone by mere narration. During my first and second expeditions Mr. Stabler was the secretary of our legation in Quito, Ecuador, where I was well acquainted. In fact for my protection in case I should find the treasure, I fully consulted the American Minister before starting on my second expedition, because from the results of my first expedition I felt sure of locating the treasure in my second trial.

My first journey was a failure because I went not sufficiently provided to stay long, and because, while climbing the "margasitas" mountain to get a look-out to see where the three Llanganati peaks were I had a fall that caused a stick to be thrust in my eye. The eye became quite irritated and swollen, although I bathed it frequently in cold water of which in those high regions there was more than an abundance. On my return to Quito the doctor prescribed hot water and an eye wash which cured it. It was the second journey when my Indian bearers deserted me---and this was when I had just pitched my camp on the edge of a lake-that proved my undoing.

This small lake lay at the base of one of the three Llanganati peaks, and since it had no visible outlet must have had a subterranean one. It was bordered on one side by a somewhat dangerous morass or swamp. I broke through the sod of this swamp and went down the full length of one leg without touching terra firma. This lake was very much like the one described in the *Derrotero* or itinerary, and since, on account of continual fogs, I had not yet seen the third and most beautiful peak (Llanganati is the Quichua or Inca word for "beautiful") I wanted carefully to examine this mysterious lake and marsh.

I had no more than constructed a rude shelter for my party when the Indians, on account of the rains and hardships, deserted me, and from these rains the lake rose gradually, finally coming up about 25 feet---way above what would have been the roof of my shelter had it remained standing instead of floating off on the surface of this little lake.

I had employed a young Ecuadorean and I had brought my servant or valet with me. These did not desert. We remained three days without fire, and after five days a relief party came out to bring me back, but on condition of not remaining in those regions.

Before this I had seen the third or most beautiful peak and found that I was within about 6 or 8 miles of it. The day that I started back was clear, and I got such a beautiful view of the third mountain that I could have cried at having to return, but I resolved to go again in a better season of the year. I had unwittingly gone in the rainy season. The seasons there in the interior were reversed, strange to say.

Regarding the *Derrotero* or itinerary of Valverde, I wish to correct the statement that I had been mystified by part of the directions in the itinerary, I had gone in there fully aware of this possible mystification, because I had read of so many people going astray by being unable to understand the diagram or hieroglyphic and make it fit the other directions. But, on account of my military education and training in map making and topography, I was conceited enough to believe that I could do better than any one before me and follow the itinerary correctly.

I did not construct a map from the *Derrotero* or guide, but I made a tracing of the map that appeared in the *Geographical Journal* of the Royal Geographical Society of London about March 1860. By the way, this map is by no means according to scale--in one day I walked what was represented by about 4 inches on the map. The next day, going just as far, I covered only 1 inch on the map.

I had no trouble in finding all of the landmarks of the guide up to and beyond the "Way of the Inca," and I had no trouble about the correct side on which to pass the "mountain which is all margasitas." By the way, this proved to be one of the 'three peaks' mentioned in the *Derrotero*.

The *Derrotero* or guide was for the purpose of guiding a party over the most practicable way of reaching the three peaks. I reached two of these peaks and plainly saw the third from its church spire-like top to its very base. This third---the beautiful mountain---is somewhat like Trinity Church with a saddle in the roof and this part snow covered, but the peak or tower mica schist, and, black, is so steep that snow cannot lie on it.

Let me make this clear and emphatic that the *Valverde Derrotero* or guide is correct and that the "three Llanganati peaks in the form of a triangle" do exist; and that a person in good health and of ordinary strength can reach them if properly provisioned and properly guided.

It took me two journeys to find the right way, but now I do not need the *Derrotero*. Those who wish to see part of the map from which I made my tracing will find this and also the full *Derrotero* in the appendix to the second volume of *Notes of a Botanist of the Ecuadorean Andes*, by Richard Spruce, a copy of which is in the New York Public Library.

E. C. BROOKS

[]

# EPILOGUE

The future cast of characters for the Llanganati Treasure story reads like a who's who list from England, Ecuador and the United States. Characters from William (The Buccaneer) Dampier, Albert Gallatin, Lewis and Clark, Richard Spruce Esq., Jordan Herbert Stabler, Colonel Edwards Cranston Brooks, Captain Eric Erskine Loch, Richard D'Orsay, Commander George Miller Dyott and Eugene Konrad Brunner among others, through Ecuador's modern day Generals Gribaldo Miño, Reñe Vargas, Frank Vargas, Medardo Salazar, Marcelo Delgado, and Presidents Roldós, Hurtado and Cordero, will all become intertwined in this intriguing story's time-line.

Much has been written concerning the Treasure of the Incas since the publication of Spruce's original paper. *Four Years Among Spanish Americans (1868)* by Friedrich Hassaurek was probably one of the first subsequent works, followed by Spruce's *Notes of a Botanist on the Amazon & Andes (1908)* published posthumously by Alfred Russel Wallace.

The list of published works on this topic flourished in the twentieth century and included; *Fever, Famine, and Gold (1938)* by Eric Erskine Loch, *Buried Gold and Anacondas (1955)* by Rolf Blomberg, *inca gold find it if you can touch it if you dare (1968)* by Jane Dolinger, *Beyond the Ranges: Five Years in the Life of Hamish MacInnes (1984)* by Hamish MacInnes, *Sweat of the Sun, Tears of the Moon (1991)* and *Lost Treasure of the Inca (1999)* by Peter Lourie, *Valverde's Gold (2004)* by Mark Honigsbaum, and three works in Spanish; *Viaje a las Misteriosas Montañas de Llanganati (1937 Second Edition 1970)* by Luciano Andrade Marín, *Llanganati (2000)* by Jorge Anhalzer, and *El Tesoro en las Misteriosas Montañas de Llanganati y La Casi Increible Historia de los Ingleses* (© 1979, 1986, 1987 published 2012) by Eugene K. Brunner.

Over a fifty year period, through his research and explorations into the Llanganati Mountains based on Spruce's and others documentation and discoveries, Eugene Brunner had become a renowned expert on the Llanganati Treasure. For most of the previously listed authors, Brunner had not only been a source of information . . . but a friend!

[]

The complete Llanganati Treasure story, which includes a reprint of this book and an English translation of the aforementioned book by Eugene K. Brunner, follows the progression of people, places and events that became part of this story over the ensuing century, is related in yet another book; *Lust For Inca Gold: The Llanganati Treasure Story & Maps* (2012).

[]

Great adventurers such as Spruce, Stabler, Brooks, Andrade, Loch, D'Orsay, Dyott, Brunner and others are introduced through short biographies that explore the true qualities distinctive to each individual, the last of the true explorers of the Llanganati region! The tale of their quests for Inca gold is related through first hand accounts of Brunner and other intrepid explorers within *Lust For Inca Gold.*

Should you choose to continue your journey by reading further, you shall attain a visual image of the conditions, hardships, deprivations, successes and failures these explorers faced on their expeditions into the mysterious Llanganati Mountains. Numerous original maps and illustrations will also provide a means to trace the explorers expedition routes as you travel with them into the unknown and witness their insatiable desire for knowledge concerning the solution to the riddle of Atahualpa's gold hidden in the Llanganatis . . . the thrill of discovering the unknown . . . a desire that drove these personal quests, or in some cases . . . their lust for Inca gold!

For the first time ever Brunner's maps and notes have been analyzed and reproduced for posterity! Brunner's writings relating to the Llanganati Treasure have also been translated and are dispersed verbatim throughout the narrative in *Lust For Inca Gold*! The totality of Brunner's discoveries, theories and conclusions concerning the location of the Llanganati Treasure will be explained in minute detail!

The story continues to relate how Brunner was no longer searching for the Llanganati Treasure . . . ultimately and after his half century quest . . . Brunner had arrived at the recovery phase! Preparing for his final expedition, the point at which he would turn his discovery over to the Government of Ecuador, Brunner would meet his untimely death and under questionable circumstances! The complete story of which is related within *Lust For Inca Gold* for the first time.

Not being just a tale about treasure, the narrative of *Lust For Inca Gold* continues to provide a clear picture of military and political conditions and events in Ecuador as they relate to the story.

Ecuador's transition from military to civilian rule becomes the setting for the conclusion of the story in modern times. Military structure and civilian administrations within Ecuador during 1972-1986 are explained and new characters are introduced. During this period coup d'états occur while others are thwarted, rivalries within the military are brought to light, the untimely death under questionable circumstances of Ecuador's President Roldós occurs, there is a mass forced retirement of Generals, yet another coup attempt is thwarted and the kidnapping of Ecuador's President Cordero transpires.

The story of Brunner's involvement with the military during this period in Ecuador transitions into a relation of events that explain this authors involvement with Brunner, the Llanganati Treasure story, Ecuador's military, and government. It is in *Lust For Inca Gold* where an intriguing little known tale of research, negotiation, romance, deceit, lies, misinformation, misdirection, betrayal, embezzlement, theft, smuggling, kidnapping and possible murder . . . is unveiled for the first time!

Most recent authors on this topic have taken to a writing style positioning themself as central characters, which also makes for interesting and easy reading. However, over the years this literary license, hearsay, misinformation and misdirection has taken hold of the story, in my humble opinion causing recent works to read more like novels than nonfiction. In truth the Llanganati story has numerous central characters, countless supporting characters and a few essential characters.

My personal involvement with this story was initially brought to light within Peter Lourie's book *Sweat of the Sun, Tears of the Moon (1991)* which was dedicated to Eugene Brunner. In *Lust For Inca Gold* I reference and critique Peter's book and *Valverde's Gold (2004)* by Mark Honigsbaum throughout, comparing, correcting and dispelling erroneous factual representations while disclosing new evidence.

Responding to my request to reprint excerpts from *Sweat of the Sun, Tears of the Moon,* Lourie himself posted on the Facebook page of *Lust For Inca Gold* . . .

> "Way to go, Steven. Finally a book that puts it all together."

Other books on this subject written in the travel adventure style prevalent in the fifties and sixties also make for interesting reading but lack substance. Sadly, the historical and substantive books on the topic have become outdated and difficult to access. *Lust For Inca Gold* fills that void, making the historical, technical and scientific knowledge on the topic, including the works of Spruce, Wallace, Stabler, Brooks, Andrade, Loch and Brunner, more accessible to the average reader.

An <u>essential character</u> to the Llanganati and Brunner story would later write concerning *Lust For Inca Gold* . . .

> "I read your book and I want to personally commend you on your perseverance and determination to undertake the painstaking task of separating fact from fiction in light of the mountain of material you had to

sift through, research, fact check, and translate. While reading your book I was amazed at how you brought all the different facets of the story together in a way that filled in the blanks and gaps. Most people that will read your book will be surprised to discover that it is so much more than a treasure hunting story. It is a history book, a snapshot of the changing politics in Ecuador at the time, and a 'must read' for any high school or university student researching any of the explorers that took part in this endeavor. I would venture to say that your book is the most detailed and accurate account of the Llanganati Treasure that has ever been put to print, and will become the definitive source of information on this subject in the future."

*Lust for Inca Gold* will be your personal expedition into the mysterious Mountains of Llanganati. Relive the history, follow the expeditions and share in the discoveries of those who have gone before you. In the end . . . from the comfort of your home . . . you will comprehend why . . . and where . . . Atahualpa's Treasure still lays hidden on Cerro Hermoso, a mountain deep within the Llanganatis of Ecuador!

Who knows . . . you may also be compelled by the more adventurous spirit within, to analyze the knowledge attained over the past century and commence your own personal quest for this enormous, yet unrecovered treasure!

###